Supplement
Nº 12.

Principles of Neurotransmission

Journal of Neural Transmission

Supplementum XII

Principles of
Neurotransmission

Proceedings of the International Symposium
of the Austrian Society for Electron Microscopy
in Cooperation with the Austrian Society for Neuropathology,
the Austrian Society for Neurovegetative Research,
and the Austrian Society for Pathology

Vienna, November 30, 1973
Edited by L. Stockinger

Springer-Verlag
Wien   New York

Univ.-Prof. Dr. Leopold Stockinger
Institut für Mikromorphologie und Elektronenmikroskopie
Universität Wien, Austria

With 61 Figures

Library of Congress Cataloging in Publication Data. Main entry under title: Principles of
neurotransmission (Journal of neural transmission: Supplementum; 12). Includes biblio-
graphies and index. 1. Neural transmission-Congresses. I. Stockinger, L., 1919—. ed.
II. Österreichische Gesellschaft für Elektronenmikroskopie. III. Series. QP363.P74. 596'.01'88.
75-2395.

ISBN 3-211-81277-6 Springer-Verlag Wien-New York
ISBN 0-387-81277-6 Springer-Verlag New York-Wien

Introduction

During the last annual meeting of the Austrian Society for Electron Microscopy it was decided to organize scientific symposia dealing with basic problems of cytology. If possible such symposia should be arranged in cooperation with other scientific societies. We intended to arrange symposia on special fields, to be presented on several aspects by excellent scientists invited.

"Neural Transmission" was chosen as a central topic for this first symposium. The subject represents a most rapidly expanding field of research and is of particular interest to morphologists, physiologists, pharmacologists, biochemists and clinicians.

We are indebted to all the invited speakers for contributing their interesting papers.

I feel bound to thank once more the firm of Bender, a subsidiary company of Boehringer Ingelheim and of Arzneimittelforschung Ges. m. b. H. Wien, for the generous financial support which made possible to realize the symposium.

The publication of this volume, representing the papers and discussions originally read at the symposium and elaborated by the authors in detail and more extensively, was only made possible by the generous cooperation of the Springer-Verlag Vienna.

Vienna, December 1974 **L. Stockinger**

Contents

Journal of Neural Transmission, Suppl. XII, 1—37 (1975)

Morphological Criteria for the Differentiation of Synapses in Vertebrates

K. H. Andres

Institut für Anatomie II, Ruhr-Universität Bochum, Federal Republic of Germany

With 22 Figures

The study of synapses is a major area of research in the field of neurobiology and neuropharmacology today. According to their function the synapses are considered by physiologists to be either electrical or chemical (Fig. 1). The chemical synapses have either an excitatory or inhibitory function. The various nomenclatures have as a common basis the connections between the different systems, for example sensory-neuronal (Figs. 2, 3), interneuronal (Fig. 1), neuro-muscular, neuroglandular and neurohumoral (Fig. 4). Since in the neurohumoral synapse a direct contact does not exist between the nerve cell and the effector system, *Scharrer* (1969 a) has suggested that the term synapsoid connection should be used instead of synapses. Naturally the chemical synapses are the most important in the area of neurochemistry and neuropharmacology and are differentiated according to the nature of the transmitter substances.

The presently known transmitter substances are acetylcholine, noradrenaline, dopamine, serotonine, GABA, glycine, glutamate, and specific peptides. With the development of better research techniques it is to be expected that other transmitter substances will be found. The transmitter substances acetylcholine and nor-adrenaline have been known for a long time and synapses of these types are referred to as cholinergic and adrenergic synapses respectively. Terminology such as aminergic, GABA-ergic has also been suggested. *Bargmann et al.* (1967) proposed the term peptidergic substances for the neurohumoral transmitters consisting of peptides occurring especially in the pars intermedia of the hypophysis.

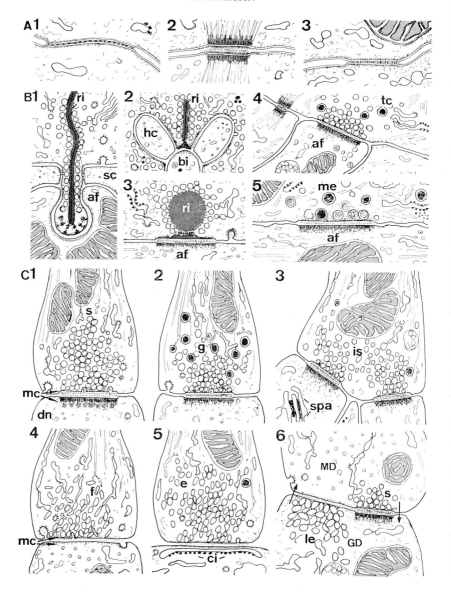

The definitions mentioned above are not sufficient to clarify the phenomena of control and modulation in the central nervous system as for example the fact that the location of a synapse on the surface of a nerve cell decides its function (Fig. 5). That is, the effect of an inhibitory, possibly GABA-ergic synapse differs depending on whether it is located on a distal dendritic section, the soma or the axon hillock. Very striking is the importance of the morphological point of view for the definition of pre- and postsynaptic inhibition. In this way the anatomical nomenclature stems from the fact that different parts of neurons are in contact with one another, in which firstly the pre- and secondly the postsynaptic sections are defined. Thus, an axo-dendritic synapse is a connection between an axon ending and a nerve cell dendrite. In a similar manner one speaks of axo-somatic, axo-axonal, somato-axonic, dendro-axonal, dendro-dendritic and somato-dendritic synapses. The frequency and distribution of the different types in the central and peripheral nervous system is extremely variable.

Apart from the topographical aspects it is the fine structure of the synaptic connection which is of special interest. In the case of electrical or electrotonic synapses the pre- and postsynaptic membranes lie in close contact with one another through a special bridge system (*Payton et al.,* 1969). In the chemical synapses there are intercellular spaces between the pre- and postsynaptic membranes. These synaptic clefts can vary in size and contain different structures. In the

---

Fig. 1. Diagrams of cellular junctions and synapses of the nervous system as revealed by electron micrographs. A 1, tight junctions between glial cell processes; A 2, desmosome between astrocyte cell processes; A 3, gap junction (electrotonic synapse) between two horizontal cells of the shark retina. B, sensoneuronal synapses with afferent nerve fibres (af); B 1, ribbon (ri) synapse of the ampulla of Lorenzini; in this case the cross section of the ribbon is about 4 to 5 microns in length; Schwann cell processes (sc). B 2, retinal cone pedicle having synaptic contact with horizontal cell processes (hc). B 3, ribbon synapse of an inner ear hair cell from caiman. B 4, taste cell (tc) synapse of rabbit foliate papillae. B 5, Merkel touch cell (me) synapse of the pig snout. C, interneuronal synapses; C 1, C 2, and C 6 represent synapses with asymmetric membrane complexes (mc) and different types of vesicle populations: type-s terminal with dominant spherical vesicles, type-g terminal with numerous granulated dense core vesicles, type-is terminal with spherical vesicles irregular in size. C 4, C 5, and C 6 represent synapses with a symmetric membrane complex and presynaptic terminals characterized by different vesicle forms: C 4, type-f terminal with numerous flattened and oval vesicles; C 5, type-e terminal with dominant oval vesicles; C 6, type-le terminal with large prominent elliptic vesicles. The arrangement of two membrane complexes in C 6 represents a reciprocal (bipolar) synapse of the olfactory bulb between a mitral cell dendrite (MD) and an external (periglomerular) granular cell (GD). Spine apparatus (spa); subsynaptic cistern of ergastoplasm (ci)

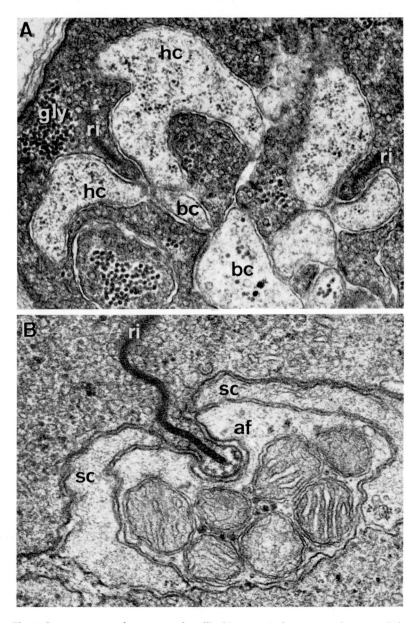

Fig. 2. Sensory neuronal synapses of scylliorhinus canicula. A, retinal cone pedicle. Presynaptic endings filled with presynaptic vesicles and glycogen granules (gly). The ribbon (ri) is in contact with the horizontal cell processes (hc) and a bipolar cell process (bc). B, ribbon (ri) contact of an electro-receptor cell of ampulla of Lorenzini to two Schwann cell processes (sc) and an afferent nerve fibre (af). The cross sectioned ribbon is about 4 to 5 microns in length. ×55,000

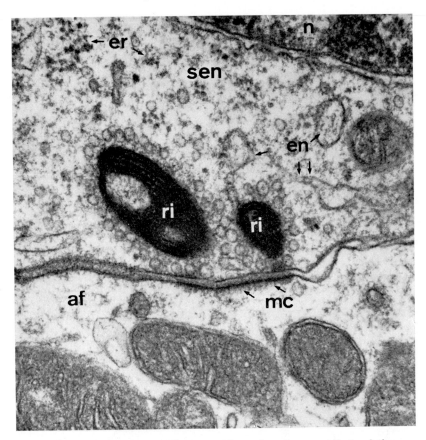

Fig. 3. Sensory neuronal hair cell synapse from the crista ampullaris of the cat. The scale formed ribbon (ri) of the type I hair cell (sen) is surrounded by synaptic vesicles which seem to be synthesized between structural elements of the nucleus (n), the ergastoplasmic reticulum (er) and endoplasmic reticulum (en). The postsynaptic dense material in the afferent nerve fibre (af) is a part of the asymmetric membrane complex (mc). ×65,000

PNS numerous synaptic endings do not directly contact the surface of the effector cell but the surrounding basal lamella (Fig. 4). A well known example of this type of synapse with the basement lamellae inserted in the synaptic clefts are the neuromuscular junctions. There are four types of neuromuscular synapses which are easily distinguished. The plate-endings on isotonic twitch fibers have subsynaptic folds (*Couteaux*, 1958). Their folds can be compared with identical structures in myotendon connections and have an adhesive function. The neuromuscular trail-endings on the isometric muscle fibers do not

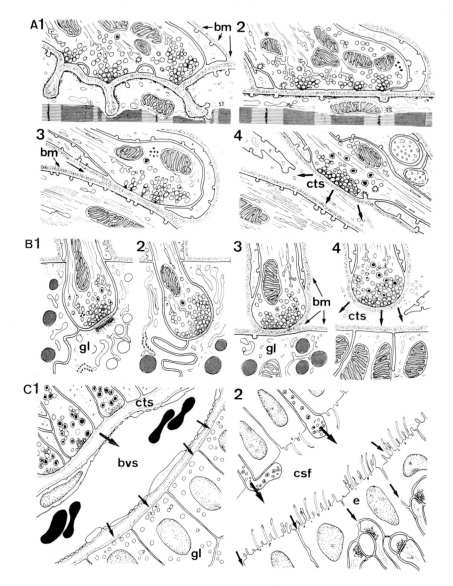

establish subsynaptic folds. Visceromotoric myoneural synapses on rapidly acting smooth muscles have similar contacts, while in slowly acting smooth muscles the synapses could be described as remote (en distance) synapses, because there are small connective tissue spaces present between the endings and the effector cells.

In neuroglandular synapses four different models of connections between nerve fibers and gland cells may be distinguished. Two types may be compared to interneuronal synapses, namely with or without postsynaptic "membrane thickenings" (see below). The other two types resemble the visceromotoric myoneural synapses (Fig. 4). Among these only the remote synapse may be considered as a transitional form leading to the neuroendocrine synapses which use the circulatory system or cerebrospinal fluid as a means of transport of the transmitter or neurohormones. The different width of the synaptic cleft, its structural organization and the topographic distribution of the receptors on the membrane of the effector cell may in addition to possessing biochemical properties be responsible for different effects caused by the same drug.

In addition to the synapses of the peripheral nervous system mentioned above the attempt to classify the synapses in the central nervous system requires more and different aspects and criteria. This article will therefore consider especially interneural synapses in the central nervous system and deal with the following questions: 1. Are there specific structures which could be correlated to the different stages in the mode of action of chemical transmitters, especially their synthesis, transport, storage, release, inactivation and uptake or resynthesis of the transmitter substance? 2. Are there structures present in the presynaptic endings that are specific for different transmitter substances? 3. Is it possible from morphological criteria to identify synapses with a specific function and could serial sections be used to reconstruct circuit diagrams? This article will also

---

Fig. 4. A, neuromuscular junction with intercalated basement lamella (bm). A 1, plate (en plaque) ending with subsynaptic folds on a frog muscle fibre; A 2, trail (en grappe) ending on an intrafusal muscle fibre; A 3, vegetative motor ending on a bronchial muscle cell; A 4, vegetative motor ending on a smooth muscle cell, remote type (synapse en distance) with a narrow connective tissue space (cts) in between. B, different types of neuroglandular synapses. B 1, asymmetric membrane complex between a nerve ending and a gland cell (gl); B 2, symmetric synapse between a nerve terminal and a gland cell; B 3, synapse with intercalated basement lamellae (bm); B 4, remote neuroglandular synapse (synapse en distance). C, neuro-humoral synapses. C 1, neurovascular chain or pathway via the blood vessel system (bvs); C 2, neuroventricular pathway via the cerebrospinal fluid (csf), and the intercellular clefts of the ependyma (e)

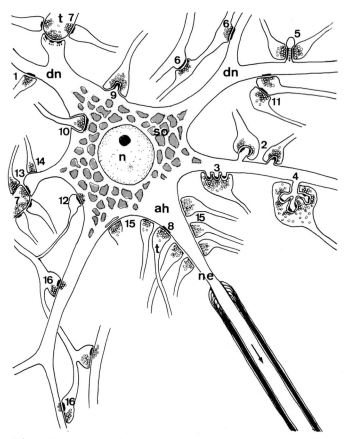

Fig. 5. Schematic representation of different types of interneuronal synapses arranged on the surface of one neuron of central nervous system; so, soma; n, nucleus; dn, dendrite; ah, axon hilloc; t, presynaptic terminal; ne, neurite. Numbers 1 to 12 indicate synapses with asymmetric membrane complexes having predominant excitatory functions. Numbers 1 to 7, axodendritic synapses; 2, spine synapse; 3, interdigitated spine synapse; 4, branched spine synapse; 5, crest synapse; 6, en passant synapse; 7, polysynaptic terminal; 8 to 11, axosomatic synapses; 9, axosomatic spine synapse; 10, invaginated, 11, axoaxonal connection; 12, somato dendritic. Numbers 13 to 16 indicate synapses with symmetric membrane complexes having a predominant inhibitory function. Presynaptic inhibition: 13, axoaxonal; 14, axodendritic. Postsynaptic inhibition: 15, axosomatic. Collateral inhibition or facilitation: 16, dendrodendritic or dendroaxonal connections of the reciprocal synapses

consider the comparative anatomy of synapses in different vertebrates with special attention to the constancy of the different structures and their specificity for corresponding functions. In this way the validity of the morphological criteria is established.

Structure of the Transmitter System

The transmitter system comprises the structural elements of the nerve cells which are required for the synthesis, transport, storage, release, inactivation and uptake or resynthesis of the transmitter substance. This secretory system is comparable both morphologically as well as functionally to a secretory cell. The regulatory mechanism controlling the release of transmitter substance is dependent on the discharge frequence of the neuron.

Secretions which are stored in the vesicles are released into the synaptic space through the process of exocytosis (*De Robertis*, 1964) in a manner similar to the release of glandular secretions. Since the site of release in the nerve cell generally is far removed from the cell body, the question of the site of synthesis of the transmitter substance arises.

From classical light microscopy, the perikaryon with its known cell organelles such as nucleus, Nissl substance, Golgi apparatus and mitochondria have been suggested as the synthetic site of the transmitter substance and not the structureless axon. The theory was supported by the discovery of axoplasma transport by *Weiss* and *Hiscoe* (1948). Criticisms of this theory are based on quantitative considerations. For example, the perikaryon of an anterior motor neuron would have to deliver acetylcholine to up to 10^4 motor endplates while in the case of some adrenergic neurons in the CNS the perikaryon might have to deliver transmitter substance to 5×10^5 endings. These quantitative considerations combined with knowledge of the ultrastructural organization of the perikaryon and peripheral sections make the reinvestigation of the localization of the transmitter synthesis necessary. Included among the transmitter substances is a group of peptide neurohormones whose synthesis depends on the ergastoplasm of the perikaryon. It is well known that in these cases the presence of storage granules with dense contents in the axon indicate the transport of these substances to the releasing site. An accumulation of deposited granules occurs in these endings. An extreme case in this regard are the Herring bodies of the neurohypophysis.

On the contrary, the biosynthesis of the relatively simple transmitter molecules acetylcholine and noradrenaline does not require the complexity of the ergastoplasm in the perikaryon. From this point of view, an axoplasmatic transport of ACh or noradrenaline is not required. The enzymes, however, which are necessary for the synthesis of the transmitter substance must be transported via the axoplasm from the perikaryon (*Sotelo* and *Taxi*, 1973; *Thoenen et al.*, 1973). Furthermore, regarding quantitative aspects a sufficient axoplasmic

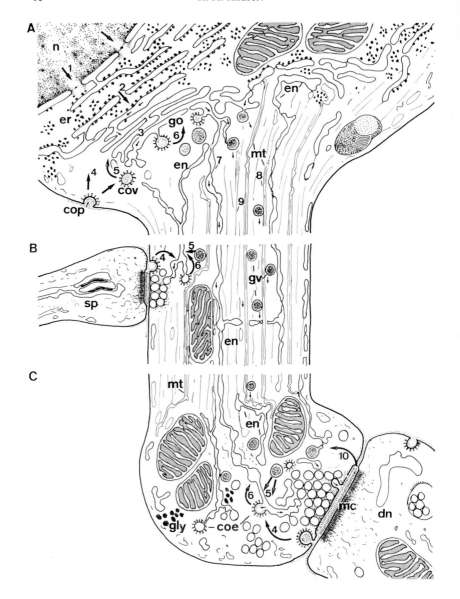

transport seems in this case possible, because the amount of enzymes involved in peripheral transmitter synthesis is relatively small. Whether they are transported via the microtubules (*Andres,* 1964), via deposited granules or via the axoplasmatic reticulum remains open to question. *Dahlström* (1968) has shown from experimental studies that following destruction of the microtubules the directed axoplasmic transport comes to a standstill. Thus, the transport of deposited granules is also inhibited.

In some cases the proven capacity of the endings to reabsorb transmitter substances or their precursor (*Whitby et al.,* 1961; *Iversen,* 1963; *Collier* and *Mac Intosh,* 1969) shows not only a rationalization with regard to the synthesis and transport capacity, but also a temporal independence of the synapse from its perikaryon.

From ultrastructural studies the hypothesis that the nerve cell process as well as the perikaryon could be necessary for a series of important metabolic functions, is supported by the fact that the mitochondria, the energy source for synthesis in the cell are especially abundant in the presynaptic section. Furthermore, the structural elements on which synthetic processes could occur *i.e.* endoplasmatic reticulum are present in all sections of the neuron (Fig. 6). Coated micropinocytotic vesicles and pits which could be regarded as morphological criteria of a resorption process (*Roth* and *Porter,* 1964; *Friend* and *Farquhar,* 1967; *Anderson,* 1969) are themselves apparent in the presynaptic endings far away from the perikaryon (*Andres,*

Fig. 6. Diagram depicting the postulated stages in the synthesis of neurotransmitter substance. A, within the perikaryon, synthesis of enzyme proteins (2) by messenger RNA (1) within the ergastoplasm (er). Completion and storage of specific enzymes in the Golgi-apparatus (3). Macromolecular information from outside of the cell body (4) and stored enzymes (5) are used for the control of enzyme and transmitter synthesis. The reabsorption and release of macromolecules (5) is carried out by coated micropinocytotic pits (cop) and coated vesicles (cov). Granulated (dense core) vesicles (gv), microtubules (mt) and canals of the smooth endoplasmic reticulum may transport neurotransmitter enzymes or substance via the axoplasm to the terminal. B, section of the neurite. The synaptic contact of a dendritic spine (sp) and chemical reabsorption (4) of macromolecular information may induce and control the synthesis of neurotransmitter within the presynaptic field of this en passant synapse of the neurite. C, synapse with an asymmetric membrane complex (mc) between a type-s terminal bouton and a dendrite (dn). Microtubules (mt) and smooth endoplasmic reticulum (en) are connected to a matrix formation in which synaptic vesicles appear. Coated pits with honeycomb like borders (coe) may transfer or control enzyme circuits (6) and the release of stored enzymes or transmitter substances (5). Number 4 indicates structures involved in the absorption of chemical information controlling the transmitter synthesis; number 10 indicates the site of reabsorption of transmitter substance and their decomposition products from the synaptic cleft respectively; glycogen granules (gly)

1964; *Andres* and *v. Düring*, 1967). This affects not only the surface membrane but also parts of the Golgi apparatus as well as formations of the endoplasmatic reticulum in the presynaptic endings, which could be similar in structure to parts of the Golgi apparatus. Vesicle like evagination of the endoplasmatic reticulum in the region of the presynaptic formations has repeatedly been described (*v. Düring*, 1967). This vesicle producing formation was most clearly shown from the endoplasmic reticulum on the smaller "en passant" synapses of the mitral cell dendrites (*Andres*, 1965) and in the corresponding axonal formations in the region of the occulomotor nuclei (*Pappas*, 1972). Fig. 6 is a schematic representation of the morphological structures which are involved in the transmitter synthesis in the different sections of the neuron.

One of these structures even more frequent in the synaptic area than in other parts of the cell is the coated pit. The function of this honeycomb like border (Fig. 6) of the coated invaginations could be more than one of pure reabsorption of macromolecular substances which have been and may be involved in reabsorption of protein in a manner similar to that observed in mosquito oocyte (*Roth* and *Porter*, 1964). It is well known that the specificity and individuality of every cell is characterized by the specific composition of their superficial glycoprotein coat (*Cook* and *Stoddart*, 1973). So, the question arises in which way the integration and orientation of the individual cell in the multicellular system of the tissue is achieved. If the surface coat should be the determinating feature every cell must have membrane systems, for the recognition of the complex macromolecular composition and the charge pattern respectively of the surface coat of the neighbouring cells. Regarding the honeycomb like border as a detector and absorption organ for the special macromolecules, then this specific coat in cells or cell parts which are particularly specialized for the recognition of other cells, must be particularly rich in coated pits. A cell which requires this mechanism is for example the macrophage as component of cellular immunity. After an experimental lesion with proton beam irradiation the macrophage appears to control the surface of Leyding cell using coated pits in deep

Fig. 7. Reaction of Leydig interstitial cells (lc) and a macrophage (mph) after proton beam irradiation. The surface coat of the Leydig cell protrusions seem to be controlled within the deep plasma membrane invaginations of the macrophage. Micropinocytotic invaginations, pits, and vesicles with coated membrane borders (cop). The cross section membrane borders seemed to be composed of very short spines. Tangential sections through the spine border elucidate its honeycomb like structure (see arrows). Intercellular space (isp); smooth and rough endoplasmic reticulum (en, er); invaginated tubules (in); lysosome (ly). ×70,000

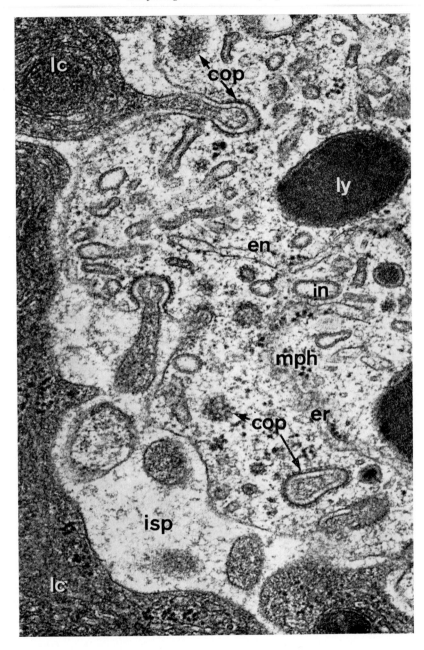

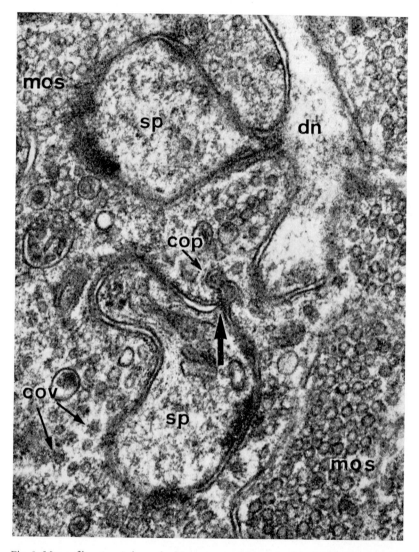

Fig. 8. Mossy fibre (mos) from the hippocampus of a dog with dendritic spines (sp). Several asymmetric synaptic membrane complexes are to be seen. Groups of small coated vesicles (cov) are intermingled between the synaptic vesicles of the mossy fibre ending. The thick arrow indicates a dark spine process invaginated into a coated pit (cop) of the presynaptic formation; dn, dendritic trunk. ×70,000

cytoplasmic invaginations (Fig. 7). In the case of complex synapses, where, besides of the coated pits, deep coated invaginations between pre- and postsynaptic membrane are frequent (Fig. 8) these invaginations may be specialized structures involved in the recognition of the postsynaptic membrane by the presynaptic process, as suggested by the striking similarity with the macrophage mentioned. Also, the small micropinocytotic vesicles with a honeycomb border occurring in the presynaptic membrane may be involved into this mechanism (Fig. 9). Perhaps this specialization of the pre- and postsynaptic membrane is additionally involved in a feedback control system regulating the synthesis of transmitter substances.

Forms of Presynaptic Endings

Different forms of presynaptic endings have already been demonstrated using the method of silver impregnation. On the basis of these results, the theory of neuronal synapses has primarily been established. The light microscopically established synapses consisting of widely distributed terminal boutons of axon branches in the CNS and the so-called gemmules of small axon branches (Fig. 1) have now been confirmed by electron microscopic studies.

In the light microscopy en passant synapses were only visible in exceptional cases in which they developed presynaptic protrusions. They are now easily recognizable in electron micrographs through the presynaptic vesicles and the membrane complexes. Telodendria with presynaptic swellings in a pearl chain arrangement may have synaptical contact to different postsynaptic processes or somata in each enlargement (Fig. 10). The presynaptic protrusions of the mossy fibers show similar arrangements. The pearl chain and the mossy fiber telodendria may be regarded as complex en passant synapses. Simple or complex en passant synapses can be established in Ranvier nodes of myelinated telodendria (*Andres*, 1965; *Pappas*, 1972).

Synaptic Vesicles

An important criterion for the differentiation of synapses is the presence of presynaptic vesicles. They are functionally interesting because they appear to store the respective transmitter substance. The investigation of certain transmitter substances associated to different kinds of synaptic vesicles is being attempted with the aid of comparative histochemistry and microautoradiographic techniques (*De Iraldi et al.*, 1963; *Whittaker*, 1965; *Hökfelt*, 1968). Until now, these techniques, with a few exceptions have had limited usage in elucidating the morphological criteria of the structure of synapses and their localization in the CNS. More attention will be given to these points in further papers in this symposium.

Differences in the form and size of synaptic vesicles could be observed. *Uchizono* (1965) described flattened vesicles as a specialized structure in some presynaptic endings on motorneurons of the spinal cord. *Andres* (1965) found four different kinds of vesicle populations in the synapses of the olfactory bulb. Beyond that, he was able to associate the described types of synapses with their origin cells. *Bodian* (1966, 1970) classified five types of vesicle populations in synapses of the spinal cord. 1. Spherical vesicles with a diameter of 450—500 Å which resist flattening after treatment with buffer, 2. flattened vesicles, 3. round or oval vesicles sensitive to flattening

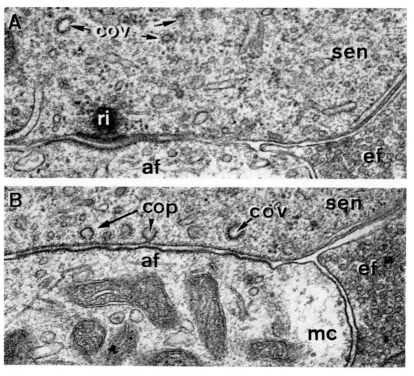

Fig. 9. Coated vesicles (cov) and coated pits (cop) of the presynaptic area in the surroundings of a synaptic ribbon (ri) of a hair cell (sen) from the papilla basilaris of the caiman. Distinct segments of smooth and granular endoplasmic reticulum are intermingled with synaptic vesicles in the presynaptic cytoplasm. Electron micrographs A and B derive from the same cell contact; afferent nerve fibre (af); efferent nerve fibre (ef) with a synaptic membrane complex (mc). ×50,000

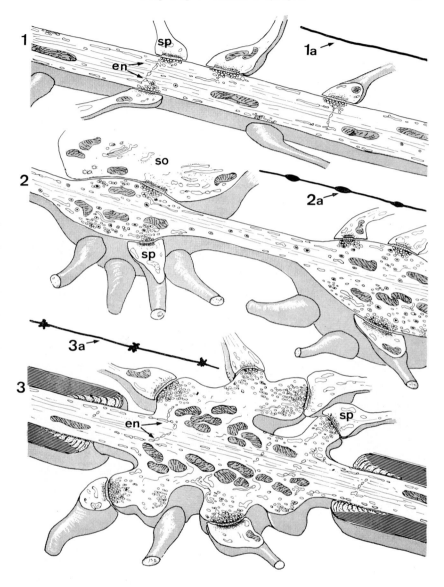

Fig. 10. Diagram of different types of axodendritic en passant synapses are revealed by electron microscopy. Golgi-preparations (1 a, 2 a, 3 a) do not show the post-synaptic spine (sp) and soma (so). 1, smooth unmyelinated presynaptic axon; 2, unmyelinated presynaptic axon with axonal swellings like a string of pearls, from the substantia gelatinosa of a dog; 3, nodal presynaptic branchlets of a myelinated mossy fibre from the cerebellum of the caiman. Smooth endoplasmic reticulum of the axoplasm (en)

18 K. H. Andres:

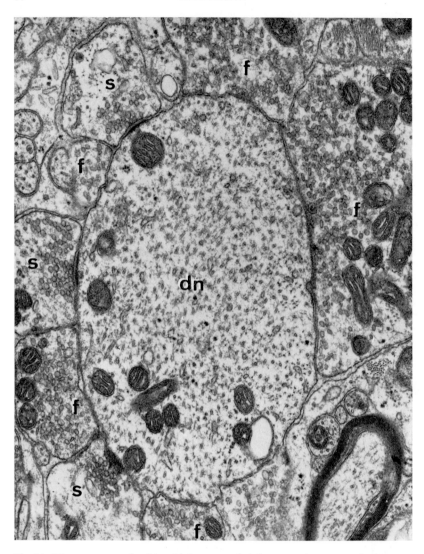

Fig. 11. Motor neuron dendrite (dn) surrounded by several presynaptic boutons from a dog. Two types of synapses may be differentiated in this section by the criteria of the structure of the membrane complex and the form of synaptic vesicles. Spherical vesicles characterize the type-s bouton and flattened vesicles are intermingled with vesicles of different size and oval forms in the type-f bouton.
×30,000

after treatment with buffer, 4. vesicles of unusually small size which are oval or flat, 5. granulated vesicles with 600—800 Å in size. In our own comparative studies on the spinal cord of several vertebrates (Figs. 11, 12) this classification of vesicle populations was confirmed with some exceptions. Even in the medulla oblongata of lamprey the types of vesicles mentioned exist (Fig. 13). After the primary optimal perfusion fixation with a glutaraldehyde formalin mixture, rinsing and storing of preparations of nerve tissue in a phosphate buffered sucrose solution (480 mosmol) at +10 °C for up to one and a half years causes no additional flattening of synaptic vesicles (Fig. 14). Using the form and size of synaptic vesicles as a criterion for the classification of synapses one has to take into consideration that there might be changes of the vesicle shape by uncommon condition of reaction or degeneration processes (Fig. 15), which occur in normal animals.

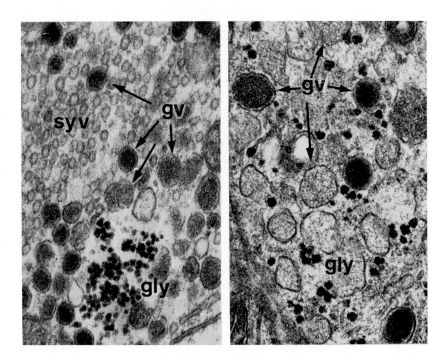

Fig. 12. Comparison of nerve terminals with granulated (dense core) vesicles (gv) from a dog. On the left: presynaptic axonal swelling (typ g) from the substantia gelatinosa rolandi (compare with Fig. 10). On the right: neuroendocrine axon from the pars intermedia of the hypophysis cerebri; gly, glycogen granules; syv, synaptic vesicles. ×55,000

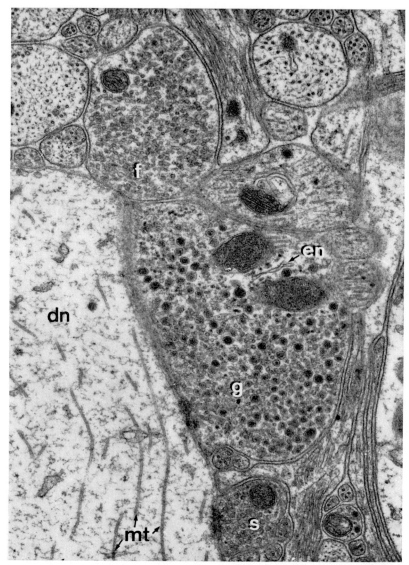

Fig. 13. The electron micrograph from the medulla of the lamprey shows identical types of synapses as the corresponding sections in the nervous system of mammals; type-f terminal with flattened vesicles (f); type-g terminal with granulated (dense core) vesicles (g); type-s terminal with spherical vesicles (s); dendrite (dn); micro-tubules (mt); endoplasmic reticulum (en). ×30,000

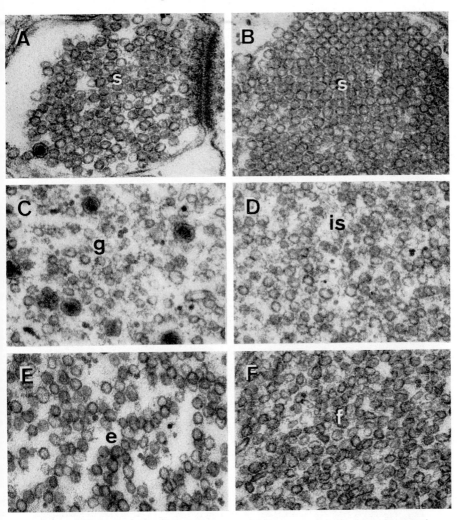

Fig. 14. Synaptic vesicles from different types of nerve terminals from the spinal cord of caiman crocodylus. The typical shape of vesicle configuration is unchanged after storage of the specimens for one and a half years in a buffered solution (see text). A, spherical vesicles (s); B, spherical vesicles forming an orderly array; C, terminal with a numerous dense core vesicles (g); D, spherical vesicles irregular in size (is); E, large elliptic synaptic vesicles (e); F, oval and numerous flattened vesicles (f). ×55,000

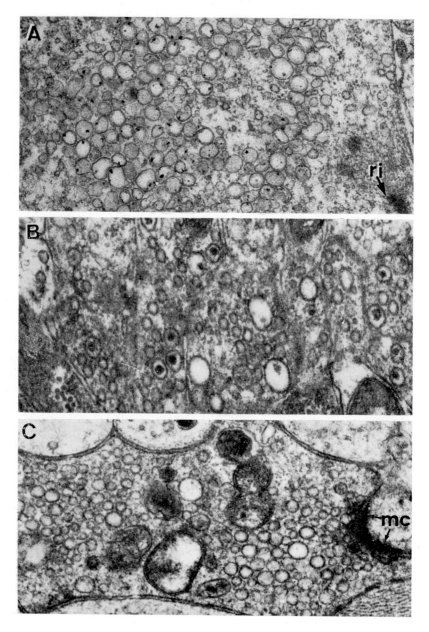

Fig. 15. Degenerative swellings of synaptic vesicles. A, presynaptic segment of a sensory hair cell of the macula utriculi from a dog; synaptic ribbon (ri). B, olfactory nerve endings from the shark olfactory bulb; C, terminal from the substantia gelatinosa of the cat; synaptic membrane complex (mc). ×55,000

Synaptic Membrane Complex

The term membrane complex implies a local differentiation of the pre- and postsynaptic membranes and the synaptic cleft between them. In this space the synaptic vesicles empty their transmitter substances by exocytosis (*De Robertis*, 1961). From the presynaptic membrane dense projections arise between the synaptic vesicles (*Akert*, 1967, 1972). These projections are visible in flat section as a regular arrangement on the presynaptic membrane (Fig. 16).

The width of the membrane space (130 Å) corresponds to the width of the usual intercellular space in some synaptic types, for instance type II of *Gray* (1959) or the L synapses of *Bodian* (1966).

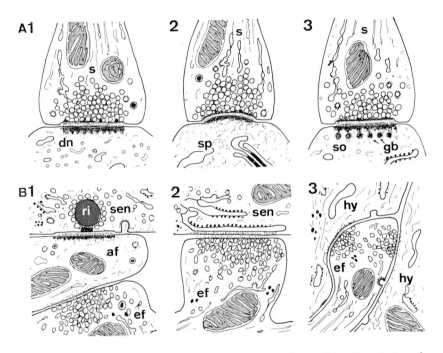

Fig. 16. Schematic representations of synapses showing the special differentiation of postsynaptic structures. A 1, postsynaptic dentated dense material in a dendritic trunk (dn). A 2, postsynaptic filamentous network of a dendritic spine (sp) containing a spine apparatus. A 3, postsynaptic dense material with globular elements (gb) from a perikaryon (so). B 1, the efferent nerve ending (ef) on the afferent sensory fibre (af) induces no postsynaptic cistern as seen in B 2 within the sensory hair cell (sen); ribbon synapse (ri). B 3, synaptic contact with hyaline cells (hy) of the organ of Corti from caiman. All synapses in A or in B may be derived from the same axon (compare Fig. 21)

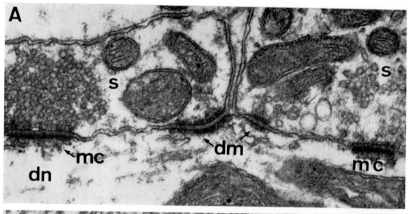

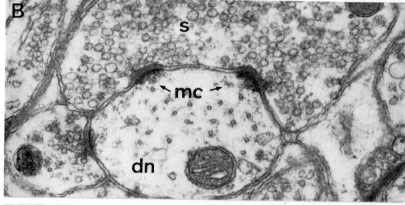

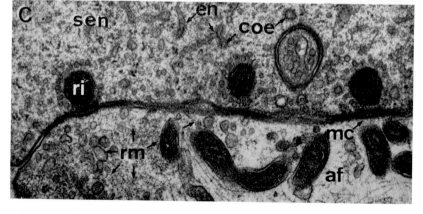

In most synapses the synaptic space is 180—200 Å and is bridged by filamentous structures with a vertical and horizontal pattern (Fig. 16). The synaptic cleft in the former type does not exhibit these structures so clearly.

The diameter of the membrane complex can be specific. In certain types of synapses the membrane complex measures one micron in diameter. The membrane complexes of most synapses have diameters of about 0.2—0.5 microns. Synaptic connections with large surface areas form as a rule more and smaller complexes (Fig. 17). In the spinal cord and in the hippocampus one finds axon terminals connected with up to 10 complexes with the corresponding dendrite (*Conradi,* 1969). Ring and horse-shoe shape or irregularly formed membrane complexes are also to be found. In sensory neuronal synapses up to five membrane complexes can exist (Figs. 17, 18).

A specialized type of synapses is the reciprocal or "bipolare" synapse (*Hirata,* 1964; *Andres,* 1965; *Rall et al.,* 1966; *Price,* 1968) which may be correlated to the physiological term of lateral inhibition, as they are found in the outer plexiform layer of the olfactory bulb of all classes of vertebrates (*Andres,* 1970) (Fig. 19) and in the inner plexiform layer of the retina between amacrine cells (*Kidd,* 1962).

A particular pattern of synaptic connections exists in the ribbon synapse of secondary sensory cells. Its presynaptic membrane complex contains a contrast rich apparatus, in which the synaptic vesicles appear to condense. In the retina this apparatus forms a ribbon like structure, in the sensory cells of electro-receptors the structures are rod shaped (Fig. 2) and in the sensory cells of lateral line system of fish as well as in the vestibulo-cochlear system they are disc or sphere shaped (Fig. 3). The light sensitive cells of the parietal eye and the related pineal cells of mammals contain ribbon synapses. Other secondary sensory cells such as those of the taste buds, the Merkel touch corpuscles as well as those cells of the Grandry corpuscles of birds have no corresponding presynaptic apparatus.

Fig. 17. Electron micrographs of synaptic contacts having more than one membrane complex (mc). A, two asymmetric synapses with desmosome junctions (dm) between the pre- and postsynaptic membranes. Desmosomes should not be mistaken for synaptic membrane complexes; dendrite (dn). ×45,000. B, terminal bouton with two membrane complexes. Serial sections have shown, that in this case the synapse has five complexes. (Compare with Fig. 18 A 3 and A 3a.) ×55,000. C, the sensory hair cell (sen) of the caiman basilar papilla has up to five ribbon synapses (ri) with one afferent nerve fibre (af). Matrix (rm) of the receptive afferent fibre; endoplasmic reticulum (en); coated vesicle (coe). ×32,000

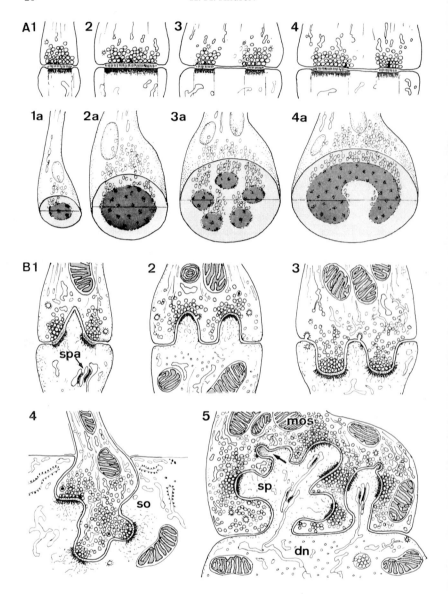

Postsynaptic Region

For the definition of a synapse the postsynaptic process is particularly important. Under the subsynaptic membranes conspicuous structural thickenings appear in different synaptic types. They can also be observed in a cross sectional view as a wide contrasting band which also can be notched. Subsynaptic globules may further characterize the pattern of the condensed substance. A fine filamentous subsynaptic material can penetrate deeper into the subsynaptic area. This subsynaptic substance is always present in the typical basic plasma of dendritic protrusions. In this protrusions subsynaptic globules are uncommon. An exception has been described by *Akert* (1967, 1972) as a crest synapse of the subfornical organ. This crest synapse is also present according to our investigations in the substantia gelatinosa Rolandi of higher mammals (Fig. 20). For the differential diagnosis of the spine synapses different modes of interdigitations between pre- and postsynaptic formations can be used. One or more fingerlike processes could penetrate relatively deep into the presynaptic terminal. In contrast to this the presynaptic formation can invade into the postsynaptic process (Figs. 8, 18). *Blackstad* (1963) could with the help of these criteria differentiate several types of mossy- and spine synapses. Their appearance is specific for certain layers of the hippocampus and the fascia dentata. Our investigations of this region of different mammals have confirmed Blackstad's findings. Similar synaptic types are to be found in different parts of the limbic system such as the amygdaloid nucleus and the septal nuclei (*Andres,* 1965). In our comparative anatomical studies of the fascia dentata we found this specially developed mossy fiber synapses (Fig. 18) exclusively in mammals.

For the differentiation of synapses it is important to know that the same presynaptic ending can have several mebrane complexes with different subsynaptic formations. Thus, the same type-s synapse with spherical vesicles can have a simple condensed subsynaptic material within a complex to a dendritic spine and a subsynaptic formation

---

Fig. 18. A, Dimensions of membrane complexes. If the area of the membrane complex exceeds more than 0.5 micron in diameter it is divided in subunits (3, 3 a) or assumes ring or horseshoe like configurations (4, 4 a). B, terminal boutons with several membrane complexes and interdigitations with the postsynaptic dendrites or spines. B 1, spine synapse with spine apparatus (spa) from the olfactory bulb. B 2, axodendritic synapse of the cerebral cortex of a snake. B 4, axosomatic synapse with an invaginated terminal from the septal nucleus of the dog, soma (so). B 5, mossy fibre synapse (mos) of the hippocampus from a rhesus monkey; spine (sp); dendrite (dn)

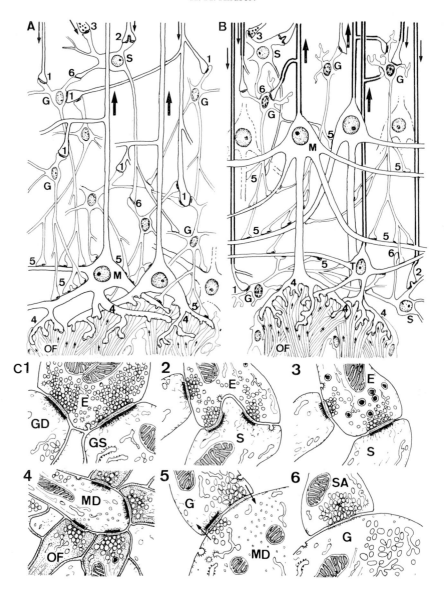

with globules within a complex to a dendrite (Fig. 21). In another case the same presynaptic fiber can have three different types of post-synaptic differentiation. Firstly, in contact with the soma of a nerve cell or sensory cell one finds a subsynaptic cisterna; secondly, if the fiber is in contact with a dendrite or sensory afferent nerve fiber only a little dense postsynaptic material is developed. Finally, in the papilla basilaris of caiman, *v. Düring* (pers. comm.) detected that the same efferent fiber which made the already described sensory-neuronal and interneuronal synaptic contacts may have synapses with supporting cells. In this case there is no postsynaptic structure to be seen (Fig. 22). From this it is clear that for the interpretation of degeneration patterns after experimental lesions by dissection or specific poisoning (*Tranzer* and *Thoenen*, 1967 , 1968) it must be taken into consideration that several types of synapses may be involved in the degeneration process even if one fiber or one fiber system was destroyed.

It is still not clear how much importance can be attached to the spine apparatus in the spine synapses for a differential diagnosis. Not every dendritic spine process contains this subsynaptic apparatus which is prominent in others.

Conclusion

Considering the described structural elements of the synapses it is possible today to differentiate between numerous types of synapses. Their classification is to be used as an instrument for the systematic analysis of circumscribed parts of the nervous system. We have

Fig. 19. Schematic representation of postulated nerve circuits in the olfactory bulb from the lamprey (A) and from a mammal (B). In both cases the same types of synapses are present in the corresponding segments. The numbers in diagrams A and B represent the types of synapses which are shown in the diagram C with ultrastructural details. The centripetal output of the bulb is indicated by large arrows. The main input into the bulb derives from the olfactory fibres (OF) (4), and from the large efferent endings (E) on the internal and external (periglomerular) granular (1) cells (G). A, further input is mediated by the stellate (short axon) cells (8) which receive centrifugal inputs via interdigitated spine synapses (2). A few granulated terminals (3) are in contact with proximal stellate cell spines. Synaptic interconnection between granular cell (G) and mitral cells (M, MO) with a bipolar arrangement (5). (6) stellate cell terminal (sa) on a granular cell process (G). Tufted cell of the external plexiform layer (T); granular cell dendrite (GD); granular cell soma (GS)

attempted to use the ultrastructure of synapses as a basis for analyzing
the circuit of neuronal connections in the rat olfactory bulb (*Andres,*
1965). Comparative studies of the olfactory bulb in several vertebrates confirmed that the method was useful (*Andres,* 1970) (Fig. 19).
The comprehension of the morphological criteria of synapses or rather
its complete characterization, can only then achieve its proper
importance when the morphological findings can be correlated with
histochemical, pharmacological and physiological results. The electron
micrograph, which mirrors actual biological events as a pale representation will become then meaningfully readable.

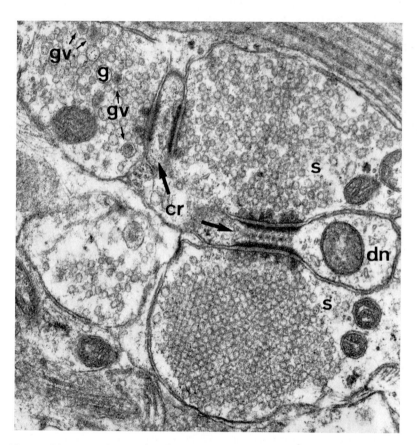

Fig. 20. Electron micrograph of two crest synapses (cr) from the substantia
gelatinosa (Rolandi) of a dog. The dendritic crests are in contact with different
presynaptic terminals. One type-s terminal shows paracrystalline arrays of vesicles.
The type-g terminal with dense core vesicles (gv) has spherical synaptic vesicles of
different size. Small dendrite (dn). ×50,000

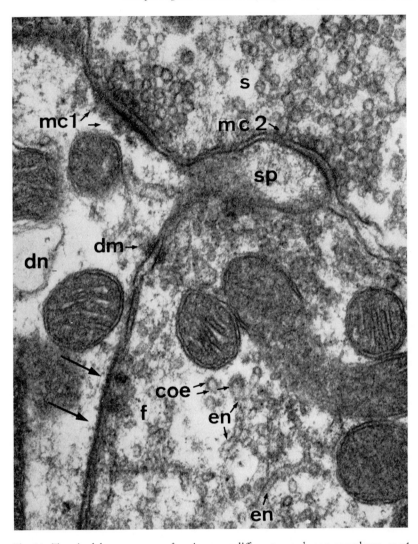

Fig. 21. Terminal bouton type-s forming two different membrane complexes, mc 1 with a dendritic trunk (dn), and mc 2 with a small dendritic spine (sp). Observe the subsynaptic globules in the case of mc 1 and the subsynaptic filamentous network in the case of mc 2. Terminal with flattened vesicles (f); smooth endoplasmic reticulum (en), coated vesicles and pits from evaginations of the reticulum (coe); arrows indicate the membrane complex of the type-f terminal; desmosome (dm).

× 65,000

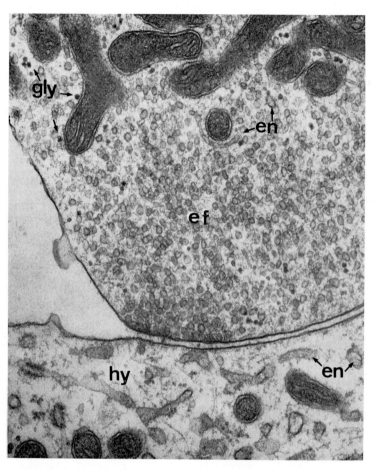

Fig. 22. Terminal of an efferent nerve fibre (ef) forming a symmetric synapse with a hyaline cell (hy) of the basal papilla of caiman crocodylus. Smooth endoplasmic reticulum (en); glycogen granules (gly). ×50,000. (Courtesy of Dr. *M. v. Düring*, 1973)

References

Akert, K., K. Pfenninger, and *C. Sandri:* The fine structure of synapses in the subfornical organ of the cat. Z. Zellforsch. *81*, 537—556 (1967).

Akert, K., K. Pfenninger, and *C. Sandri:* Crest synapses with subjunctional bodies in the subfornical organ. Brain Research 5, 118—121 (1967).

Akert, K., K. Pfenninger, and *C. Sandri:* Structure and Function of Synapses, freeze etching and cytochemistry of vesicles and membrane complexes in synapses of the central nervous system, pp. 67—88. New York: Raven Press. 1972.

Anderson, E.: Oogenesis in the cockroach, Periplaneta Americana, with special reference to the specialization of the oolemma and the fate of coated vesicles. J. Micros. *8*, 721—738 (1969).

Andersson-Cedergren, E.: Ultrastructure of motor end-plate and sarcoplasmic components of mouse skeletal muscle fiber. J. Ultrastruct. Res. Suppl. *1*, 15—180 (1959).

Andres, K. H.: Mikropinozytose im Zentralnervensystem. Z. Zellforsch. *64*, 63—73 (1964 b).

Andres, K. H.: Der Feinbau des bulbus olfactorius der Ratte unter besonderer Berücksichtigung der synaptischen Verbindungen. Z. Zellforsch. *65*, 530—561 (1965).

Andres, K. H.: Über die Feinstruktur besonderer Einrichtungen in markhaltigen Nervenfasern des Kleinhirns der Ratte. Z. Zellforsch. *65*, 701 to 712 (1965).

Andres, K. H.: Der Feinbau des Subfornikalorganes vom Hund. Z. Zellforsch. *68*, 445—473 (1965).

Andres, K. H., and *M. von Düring:* Mikropinozytose in motorischen Endplatten. Naturwissenschaften *23*, 615—616 (1966).

Andres, K. H.: Anatomy and ultrastructure of the olfactory bulb in fish, amphibia, reptiles, birds and mammals. In: Taste and Smell in Vertebrates, pp. 177—196. London: Ciba Foundation. 1970.

Andres, K. H., and *M. von Düring:* Interferenzphänomene am osmierten Präparat für die systematische elektronenmikroskopische Untersuchung. Mikroskopie (Wien) *30*, 139—149 (1974).

Bargmann, W., E. Lindner, and *K. H. Andres:* Über Synapsen an endokrinen Epithelzellen und die Definition sekretorischer Neurone. Z. Zellforsch. *77*, 282—298 (1967).

Blackstad, T.: Ultrastructural studies on the hippocampal region, pp. 122 to 148. In: The Rhinencephalon and related Structures. (Progress in Brain Research, Vol. 3.) 1963.

Bodian, D.: Synaptic types on spinal motoneurons. An electron microscopic study. Bull. Hopk. Hosp. *119*, 16—45 (1966).

Bodian, D.: An electron microscopic characterization of classes of synaptic vesicles by means of controlled aldehyde fixation. J. Cell Biol. *44*, 115 to 124 (1970).

Bodian, D.: Synaptic diversity and characterization by electron microscopy, pp. 45—65. In: Structure and Function of Synapses. New York: Raven Press. 1972.

Collier, B., and *F. C. Mac Intosh:* The source of choline for acetylcholine synthesis in a sympathetic ganglion. Can. J. Physiol. Pharm. *47*, 127 to 138 (1969).

Conradi, S.: On motoneurons synaptology in adult cats. An electron microscopic study of the structure and location of neuronal and glial elements on cat lumbosacral motoneurons in the normal state and after dorsal root section. Acta physiol. scan. Suppl. *332*, 1—115 (1969).

Conradi, S.: On motoneurons synaptology in kittens. An electron microscopic study of the structure and location of neuronal and glial elements on cat lumbosacral motoneurons in the normal state and after root section. Acta physiol. scand. Suppl. *333,* 1—76 (1969).

Cook, G. M., and *R. W. Stoddart:* Surface carbohydrates of the eukaryotic cell, p. 115. London: Academic Press. 1973.

Couteaux, R.: Morphological and cytochemical observations on the postsynaptic membrane at motor end-plates and ganglionic synapses. Exptl. Cell. Res. Suppl. *5,* 294—322 (1958).

Dahlström, A.: Effect of colchicine on transport of amine storage granules in sympathetic nerves of rat. Europ. J. Pharmacol. *5,* 113—123 (1968).

De Iraldi, A. P., H. F. Duggan, and *E. De Robertis:* Adrenergic synaptic vesicles in the aterior hypothalamus of the rat. Anat. Rec. *145,* 521 to 531 (1963).

De Robertis, E.: Ultrastructure of the synaptic region, pp. 3—48. In: Histophysiology of Synapses and Neurosecretion. Pergamon Press. 1964.

Dowling, J. E., and *B. B. Boycott:* Organization of the primate retina: electron microscopy. Proc. Roy. Cox. B *166,* 80—89 (1966).

von Düring, M.: Über die Feinstruktur der motorischen Endplatte von höheren Wirbeltieren. Z. Zellforsch. *81,* 74—90 (1967).

Friend, D. S., and *M. G. Farquhar:* Functions of coated vesicles during protein absorption in the rat vas deferens. J. Cell Biol. *35,* 357—376 (1967).

Gray, E. G.: Axo-somatic and axo-dendritic synapses of the cerebral cortex: an electron microscope study. J. Anat. (London) *93,* 420—433 (1959).

Gray, E. G.: A morphological basis for presynaptic inhibition? Nature *193,* 82 (1962).

Hirata, Y.: Some observations on the fine structure of the synapses in the olfactory bulb of the mouse, with particular reference to the atypical synaptic configuration. Arch. Histol. Japan. *24,* 293 (1964).

Hökfelt, T.: In vitro studies on central and peripheral monoamine neurons at the ultrastructural level. Z. Zellforsch. *91,* 1—74 (1968).

Iversen, L. L.: The uptake of noradrenaline by the isolated rat heart. Brit. J. Pharmacol. *21,* 523—537 (1963).

Kidd, M.: Electron microscopy of the inner plexiform layer of the retina in the cat and pigeon. J. Anat. *96,* 179—190 (1962).

Milhaud, M., and *G. D. Pappas:* The fine structure of neurons and synapses of the habenula of the cat with special reference to sub-junctional bodies. Brain Research *3,* 158—173 (1966 b).

Pappas, G. D., and *St. G. Waxman:* Synaptic fine structure—morphological correlates of chemical and electrotonic transmission, pp. 1—43. In: Structure and Function of Synapses. New York: Raven Press. 1972.

Payton, B. W., M. V. L. Benett, and *G. D. Pappas:* Permeability and structure of junctional membranes at an electrotonic synapse. Science *166,* 1641—1643 (1969).

Price, J. L.: The termination of centrifugal fibers in the olfactory bulb. Brain Research *7,* 483—486 (1968).

Rall, W., G. M. Shepherd, F. S. Reese, and *M. W. Brightman:* Dendro-dendritic synaptic pathway for inhibition in the olfactory bulb. Exptl. Neurol. *14,* 44—56 (1966).

Roth, T. F., and *K. R. Porter:* Yolk protein uptake in the oocyte of the mosquito Aedes aegypti L. J. Cell Biol. *20,* 313—332 (1964).

Scharrer, B.: Neurohumors and Neurohormones: Definitions and terminology. J. Neuro-visc. Rel., Suppl. *9,* 1—20 (1969).

Sotelo, C., and *J. Taxi:* On the axonal migration of catecholamines in constricted sciatic nerve of the rat. Z. Zellforsch. *138,* 345—370 (1973).

Weiss, P., and *H. Hiscoe:* Experiments on the mechanism of nerve growth. J. exp. Zool. *107,* 315—396 (1948); *138,* 345—370 (1973).

Whittaker, V. P.: The application of subcellular fractionation techniques to the study of brain function. Prog. Biophys. Mol. Biol. *15,* 39—48 (1965).

Whitby, L. G., J. Axelrod, and *H. Wil-Malherbe:* The fate of ^{3}H-norepinephrine in animals. J. Pharm. Exp. Ther. *132,* 193—201 (1961).

Thoenen, H., U. Otten, and *F. Oesch:* Axoplasmic transport of enzymes involved in the synthesis of noradrenaline: relationship between the rate of transport and subcellular distribution. Brain Research *62,* 471 to 475 (1973).

Tranzer, J. P., and *H. Thoenen:* Ultramorphologische Veränderungen der sympathischen Nervenendigungen der Katze nach Vorbehandlung mit 5- und 6-Hydroxy-Dopamin. Naunyn-Schmiedebergs Arch. exp. Path. Pharmak. *257,* 343—344 (1967).

Tranzer, J. P., and *H. Thoenen:* An electron microscopic study of selective, acute degeneration of sympathetic nerve terminals after administration of 6-hydroxydopamine. Experientia (Basel) *24,* 155—156 (1968).

Uchizono, K.: Characteristics of excitatory and inhibitory synapses in the central nervous system of the cat. Nature *207,* 642—643 (1965).

Author's address: Prof. Dr. *K. H. Andres,* MD, Institut für Anatomie II der Ruhr-Universität, Universitätsstraße 150, D-4630 Bochum, Federal Republic of Germany.

Discussion

Ratzenhofer: As regards Prof. Andres' classification of synapses, his naming of "neuroglandular and neurohumoral synapses" is a matter of basic concern. Literally, a *synapsis* is understood to be a site of stimulus transmission, whereby the afferent, usually varicose axon segment with its special differentiations comes into *direct contact* with the nerve cell process to be innervated or, in the case of the peripheral nervous system, with very specific target cells. This, however, is only seldom the case with peripheral nerve endings. Altogether the following three possibilities are to be found, as first suggested by our group in Graz with *Müller* and *Becker* in 1969 (see figure).

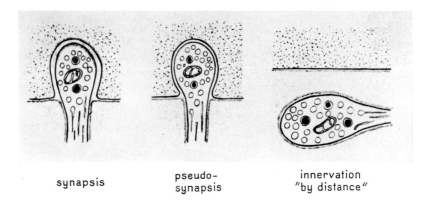

synapsis	pseudo-synapsis	innervation "by distance"

1. The *(true) synapsis* with direct contact. Example: the cells of the adrenal medulla. In this case the basal membrane covering the terminal transmitter segment extends directly into the basal membrane of the target cell.

2. A topographically very similar situation is presented by the intra-epithelial nerve endings, which have also been described by other authors: again, the basal membrane partly covers the terminal segment and, in direct continuation, turns over on the epithelial cell. The differences in comparison to the (true) synapse: even in serial sections no characteristic differentiation of the cleft, no thickening of the membranes, etc. We have thus named these nerve endings *pseudosynapses* and suspect that they are afferent nerve endings.

3. In the gastrointestinal tract as well as in vascular connective tissue and other mesenchyme the well-documented situation is usually to be found in which the nerve endings are more or less *distant* from the suspected target cell. The nerve ending is also separated from the target cell by two basal membranes and intermediate substances or, in the case of innervation of the naked connective-tissue cell by one membrane and intermediate substance, as described by Prof. *Pischinger* and *Stockinger*'s group since 1967.

We have thus named this type of neural supply *"innervation by distance"* (or *"distant innervation"*), with a certain indebtedness to *Jabonero*'s and *Ruska* and *Ruska*'s (1961) term "synapsis by distance". As this involves an intrinsic contradiction, I still prefer the designation "innervation by distance" when appropriate for the neuroglandular and neurohumoral types of the supply mentioned by the speaker. Its functional purpose is the distribution and dilution of neurohormones in tissue fluid, and the thus delayed effect and reaction of the target cells. This is reasonable in the gastrointestinal and other organ systems in which "emergency reactions" *(Cannon)* as in the adrenal medulla would be totally un-desiderable.

[Lit.: *Jabonero, V.:* Acta neuroveget. (Wien) *29*, 11 (1966); *Ratzenhofer, M., O. Müller,* and *H. Becker:* Mikroskopie *25*, 297 (1969); *Ruska, H.,* and *C. Ruska:* Dtsch. Med. Wschr. *1961*, 1697, 1770.]

Sotelo: If as some cytologists suggest, Golgi apparatus and endoplasmic reticulum are two distinct cell organelles, why do you call Golgi apparatus to the tubular and vesicular profiles of smooth endoplasmic reticulum found in some central axon terminals? Probably the hyperplasia of such endoplasmic reticular profiles, present in some of your slides, are in relationship to the axonal remodeling process we have suggested (*Sotelo* and *Palay:* Lab. Invest. 1971) to occur in axon terminals of normal-looking animals.

Andres: In my opinion portions of the smooth endoplasmic reticulum in the presynaptic endings can carry out simple synthetic functions, comparable with some of the functions of the Golgi-apparatus in the perikaryon.

Whittaker: I am a bit worried by your results suggesting the presence of endoplasmatic reticulum in nerve terminals. Warming synaptosomes in sucrose causes intracytoplasmic fusion of vesicles with the formation of tubules. So far as we know this change is irreversible. Might not the appearance of tubular membrane structures in nerve terminals be also due to an autolytic reaction during fixation?

Andres: I know that many investigators don't accept the tubular network or tubular reticulum in the presynaptic region. Reconstruction of our material from serial sections shows the smooth endoplasmatic reticulum as a regular cell structure in the ending. I don't believe that our tubules are artefacts due to autolytic reaction. Our perfusion fixation technique has been developed with consideration of such possible artefacts.

Zenker: In synaptic endings of nerve fibres you showed neurotubules in close relation to synaptic vesicles. In your schematic drawings I even observed a transition of tubules into synaptic vesicles. Do such transition actually exist? I cannot imagine it.

Andres: No. The synaptic vesicles do not bud off from the microtubuli. However, it can't be ruled out that during the alteration of the microtubuli in the presynaptic ending, the endoplasmic reticulum has a functional link with the microtubuli.

Journal of Neural Transmission, Suppl. XII, 39—60 (1975)
© by Springer-Verlag 1975

The Biochemistry of Cholinergic Synapses as Exemplified by the Electric Organ of *Torpedo*

V. P. Whittaker, H. Zimmermann, and **M. J. Dowdall**

Abteilung für Neurochemie, Max-Planck-Institut für biophysikalische Chemie, Göttingen, Federal Republic of Germany

With 17 Figures

Introduction

The electric organ of *Torpedo* (Fig. 1) consists of two masses of gelatinous tissue disposed one on each side of the head, having a honeycomb structure when viewed from above. Each "cell" of the honeycomb is in reality a vertical stack of electroplaque cells, each profusely innervated on the lower (ventral) surface; when the electric nerves discharge synchronously, the postjunctional potentials thus generated summate to give sizeable electric discharges (25—35 v in *T. marmorata*).

The transmitter at these junctions is acetylcholine; thus electric organs are an extremely rich source of cholinergic nerve terminals. The acetylcholine content of the tissue may exceed 1000 nmol/g compared with 1—2 nmol/g in the guinea pig diaphragm. Electric tissue thus provides the same kind of model synapse for the investigation of the cholinergic system as the adrenal medulla does for the noradrenergic synapse. The main difference between the presynaptic nerve terminals of electric tissue (Fig. 2) and other types of cholinergic synapse is the larger size of the vesicles — 800—900 Å as compared with 400 to 500 Å for other tissues.

The cholinergic tracts originate in prominent, readily accessible, yellowish lobes (the electric lobes) in the brain stem (Fig. 3). The perikarya of the very large electromotor neurones are bounded by a histologically cleear zone; they do not receive axosomatic synapses and the purely excitatory innervation is all axodendritic. Thus a valuable feature of the preparation is that all parts of a single, well-defined cholinergic system are accessible and in quantity: the cell bodies, the axons and the terminals in the electric tissue. In preliminary work (with *L. Fiore* and *R. J. Thompson* respectively) we have noted the feasibility of isolating the perikarya and nuclei of the electromotor neurones. The axons can be used for the study of axonal flow in a cholinergic neurone (*Zimmermann* and *Whittaker*, 1973 a; *Ulmar* and *Whittaker*, 1974 a, b; *Heilbronn* and *Peterson*, 1973; *Widlund, Karlsson, Winter* and *Heilbronn*, 1974). The terminals in the electric tissue provide us with isolated cholinergic synaptic vesicles (*Whittaker, Essman* and *Dowe*, 1972 b) and enable the effects of stimulation on the number, size and composition of synaptic vesicles to be studied (*Zimmermann* and *Whittaker*, 1973 b, 1974 a, b). As far as the postsynaptic side is concerned, the electric tissue is an excellent source of the acetylcholine receptor (*Changeux*, 1972; *Miledi, Molinoff* and *Potter*, 1971; *Schmidt* and *Raftery*, 1972; *Karlsson, Heilbronn* and *Widlund*, 1972).

In this report a brief survey of the work of our own group will be given: for further information the reader is referred to other recent reviews (*Whittaker*, 1971, 1973 a, b, 1974; *Whittaker* and *Dowdall*, 1973; *Whittaker, Dowdall* and *Boyne*, 1972 a).

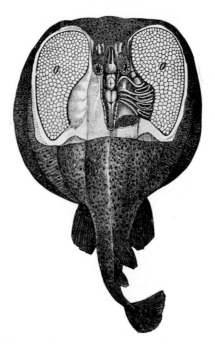

Fig. 1. The electric organs (O) of *Torpedo* (*Fritsch*, 1890)

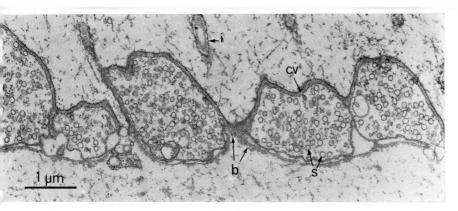

Fig. 2. Nerve terminals in electric tissue: s, synaptic vesicles within nerve terminal profile; i, invagination of postsynaptic membrane; cv, coated vesicle; b, basal membrane

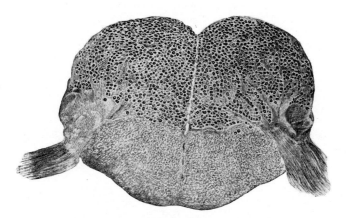

Fig. 3. Cross-section of electric lobes showing prominent perikarya of electromotor neurones (*Fritsch*, 1890)

Subcellular Fractionation of the Terminals

The Isolation and Composition of the Vesicles

Electric tissue is difficult to homogenize and does not yield synaptosomes in appreciable amounts when comminuted. We have been able to isolate vesicles on a relatively large scale (*Whittaker et al.*, 1972 b) by crushing the tissue frozen in liquid nitrogen to a

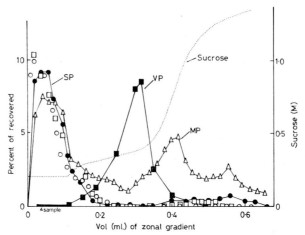

Fig. 4. Separation of acetylcholine-containing vesicles (VP) from soluble cytoplasmic protein (SP) and membrane fragments (MP). The soluble protein constituents include lactate dehydrogenase (○) choline acetyltransferase (□) and solubilized cholinesterase (△) the portion of which that remains membrane bound is recovered in the MP peak. Other symbols: protein (●); acetylcholine (▉)

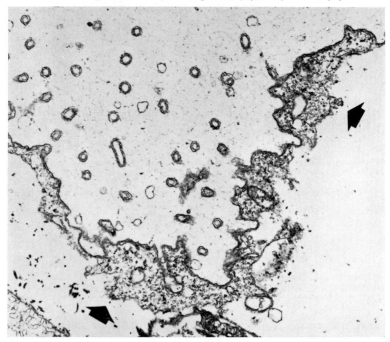

Fig. 5. Fragment of electroplaque showing (arrow) synaptic vesicles held in place by frozen cytoplasm below severely disrupted external membrane. After extraction with sucrose-NaCl such regions are empty. Glutaraldehyde-osmium-Epon

Table 1. *Comparison of Amine and ATP Contents of Various Storage Granules*

Amine	Species	Cell or tissue	Amine content (nmol/g of protein)	Amine: ATP molar ratio	References
Adrenaline	Ox	Adrenal medulla	3450	4.55	*Smith* (1968), *Hillarp* (1958)
Noradrenaline	Rat	Stellate ganglion	—	3.12	*Geffen* and *Livett* (1971),
		Heart	—	2.9—4.2	*Livett* (1971),
		Vas deferens	—	1.3—3.7	**Schümann* (1958)
	Ox	Stellate ganglion	—	3.12	
		Splenic nerve	9	3.0—5.2* / 7.5—12.0+	*de Potter et al.* (1970)
Serotonin + Histamine	Rabbit	Platelet	210 / 90	3.5	*Da Prada* and *Pletscher* (1968)
Acetylcholine	*Torpedo*	Electric organ nerve terminals	680—1200	5.3	*Dowdall et al.* (1974)

+ Allowance made for ATP contributed by mitochondrial contamination.

coarse powder and extracting it with 0.2 M sucrose—0.3 M NaCl: after removal of large and intermediate-sized particles by centrifuging (10,000 g × 30 min) the supernatant (the "cytoplasmic extract") is separated (Fig. 4) on a density gradient into three peaks comprising soluble cytoplasmic proteins (SP), vesicles (VP) and membrane fragments (MP). Examination of the tissue fragments thus extracted in the electron microscope reveals that extraction of vesicles has been facilitated by the extensive disruption of the external presynaptic membrane away from the synaptic cleft (*Soifer* and *Whittaker*, 1972) which occurred during crushing (Fig. 5).

The VP peak is an almost pure preparation of vesicles (Fig. 6). Analysis has shown that it contains four main protein constituents, three of which are located in the membrane and the fourth of which is isolated as a low-molecular weight acidic protein (m.wt. 10,000, isoelectric point $\simeq$ 3.0) and is believed to be derived from the core. This last (vesiculin) is associated with a nucleotide, probably AMP, which we believe is derived during isolation from ATP, also present in the vesicles (Figs. 7, 8) (*Dowdall, Boyne* and *Whittaker*, 1974). The composition of the vesicles (biogenic amine stored in combination with an acidic core protein and ATP) thus resembles other storage granules (Table 1) (*Whittaker, Dowdall, Dowe, Facino* and *Scotto*, 1974).

Effect of Stimulation on Vesicle Yield and Composition

The effect of electrical stimulation of the electromotor neurones on the yield and composition of the vesicles and the morphology of the terminal has been studied after stimulation (*Zimmermann* and *Whittaker*, 1974 a) and recovery (*Zimmermann* and *Whittaker*, 1974 b). Stimulation was applied directly to the electromotor nucleus in the brain stem after cutting the electromotor nerves on one side so as to preserve a non-stimulated control organ. After varying periods of stimulation, samples were taken from both sides for electron microscopy and vesicles were isolated from the excised and frozen organs.

Morphological Changes

On stimulation there is a progressive fall in both vesicle numbers and size as shown by electron micrographs of whole tissue (Figs. 9, 10). At the same time, the external presynaptic membrane increases in area; blebs and infoldings appear in it, with the result that the presynaptic terminal profiles become more numerous and smaller. These changes are consistent with the loss, by exocytosis, of about half the vesicle population. On recovery, the number and size of the vesicles are back to normal within 24 hours; as explained below, this is well before complete physiological or biochemical recovery takes place.

Biochemical Changes

The decline in vesicular protein and total nucleotide as measured by comparing the vesicular fractions isolated from stimulated organs with those isolated from unstimulated controls is in reasonable agreement with the decline in vesicle numbers as measured morphologically (Fig. 9). This implies that both the vesicle membrane and the core (as far as protein and total nucleotide is concerned) of the surviving vesicles retain their integrity. However, these vesicles are on average depleted both in transmitter and ATP (Fig. 9). On recovery, acetylcholine and ATP levels remain depressed for up to 2 days; thereafter acetylcholine rises to normal or even to supernormal levels, but vesicular ATP remains depressed as long as 8 days after

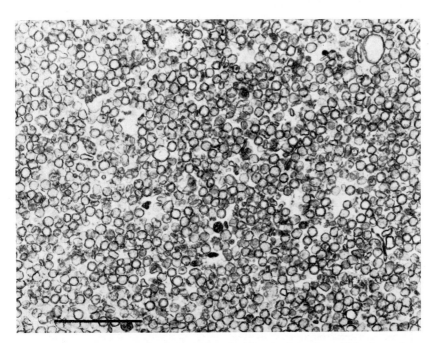

Fig. 6. Electron micrograph of the VP peak (Fig. 4) after glutaraldehyde-osmium fixation followed by centrifugation, dehydration of the pellet and embedding in Epon. Bar is 1 μm

prolonged stimulation (Fig. 11). Restimulation during recovery again causes vesicle depletion, indicating lost of transmitter-free vesicles.

A study of the ratio of vesicular acetylcholine to ATP under different conditions of rest, stimulation and recovery has some interesting implications (Fig. 12). Vesicles isolated from control organs removed under anaesthesia may have ratios as high as 10 : 1. Vesicles isolated from organs stimulated repetitively 500—20,000 times have a lower ratio [4.03 ± 0.39 (11)]; the progressively lower amounts of vesicular acetylcholine and ATP recovered in the peak fraction are highly correlated. On recovery, although vesicular acetylcholine and ATP levels remain depressed for several days, the ratio rises again to very high ($\simeq 10 : 1$) levels.

These findings suggest that the vesicle population in a normal (resting) organ is heterogeneous and that the most labile vesicles are particularly rich in acetylcholine. Further evidence (Fig. 12) for vesicle heterogeneity comes from examining the acetylcholine: ATP

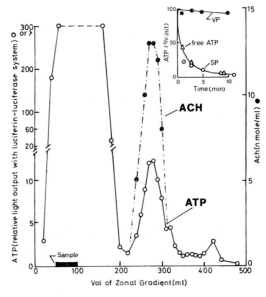

Fig. 7. Distribution of ATP in the gradient following a zonal centrifugal separation of the cytoplasmic extract of an electric organ of *Torpedo*. Note the large soluble cytoplasmic peak of free ATP (SP) which is destroyed (insert) by an apyrase-myokinase mixture at the same rate as an equivalent amount of authentic, free ATP and the smaller peak of vesicular ATP (VP), coincident with the vesicular peak of acetylcholine (ACH) and which is protected (insert) by the vesicular membrane from the action of hydrolytic enzymes

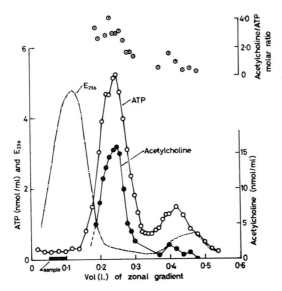

Fig. 8. An experiment similar to that shown in Fig. 7 in which the cytoplasmic extract was first treated with the apyrase-myokinase mixture before centrifuging. Notice the disappearance of the first (SP) peak and the emergence of only one main, vesicular, peak of ATP as expected from the results shown in the insert to Fig. 7

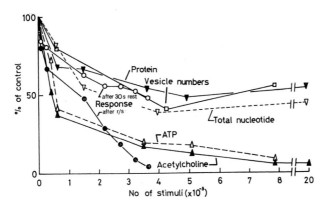

Fig. 9. Effect of stimulation on vesicle numbers, vesicular constituents (protein, total nucleotide, ATP and acetylcholine) and response of the organ (r/s) to repetitive stimuli. Points are the means of single experiments expressed as percentages of the control (unstimulated) organ, except for the electrophysiological response which is expressed as a percentage of the initial response. Note that vesicle numbers, protein and total nucleotide fall *paripassu* with the capacity of the organ to respond to single stimuli after an optimum (30 sec) rest period (white circles) whereas vesicular acetylcholine and ATP fall more profoundly

ratio across the vesicle peak. In Fig. 12-2 the ratio has been plotted for a number of experiments with organs taken from fish killed without anaesthesia in which some spontaneous stimulation had occurred. The ratios of fractions far from the peak are not very reliable as they are the ratios of two small numbers. However, there are significant differences and this suggests that the different types of vesicle may vary to some extent in density and thus be partially separable.

Physiological Studies

Fig. 9 shows the response of the organ to repetitive stimulation via the electromotor nucleus. It will be seen that the response falls rather rapidly during the initial phase of stimulation, then tends to stabilize and finally falls again to almost zero. During the period of initial fall, vesicles disappear, acetylcholine: ATP ratio falls and the levels of acetylcholine and ATP fall though not, as we have seen (Fig. 12), *paripassu*. If at this stage the capacity of the response to recover is tested with single shocks it is found that after the initial

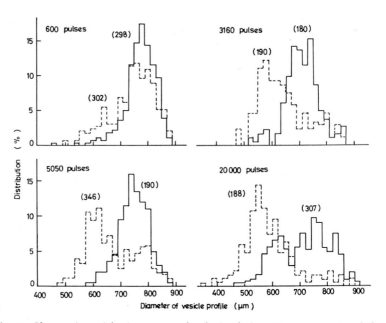

Fig. 10. Changes in vesicle size as a result of stimulation. Note progressive shift of mean vesicle profile diameter to smaller value on stimulation (broken lines). After short periods of stimulation the distribution of profile diameter is bimodal. The figures in parentheses are the number of vesicle profiles measured

fall complete recovery requires a matter of hours; short-term recovery does not exceed 50 %. However, even after prolonged stimulation with very low vesicular acetylcholine and ATP ratios, this partial short-term recovery is still observed.

We have seen that recovery of vesicle numbers proceeds much faster than recovery of vesicular acetylcholine and ATP. As vesicle numbers return to normal and the acetylcholine: ATP ratio rises to values even higher than in the unstimulated organ, the responsiveness of the organ to single shocks returns to normal and may even exceed normal values. However, as long as vesicular acetylcholine and ATP concentrations remain low, the organ is abnormally fatiguable as judged by the number of repetitive stimuli required to reduce its response to 50 % of control.

One may summarize these findings by saying that the organ behaves as though its ability to respond optimally to single shocks is a function of the vesicle content of the terminals whereas its ability to respond to repetitive stimulation is a function of the transmitter

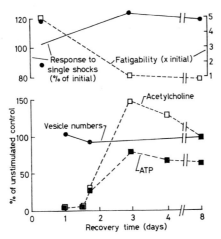

Fig. 11. Recovery of (lower diagram) vesicular acetylcholine and ATP and vesicle numbers and (upper diagram) response of the organ to single shocks following repetitive stimulation with 3500—7000 stimuli, enough (Fig. 9) to deplete vesicle numbers to 22—50 %, acetylcholine to 5—13 % and ATP to 8—21 %.

It will be seen that although the response of the organ to single shocks has recovered more or less completely by 8 hours and vesicle numbers by 24 hours, vesicular acetylcholine remains depressed for at least 2 days. The fatigability (measured by the reciprocal of the number of pulses at 5/sec required to reduce the response of the organ to mV levels) of the stimulated organ returns to normal paripassu with the restoration of normal transmitter levels and not with the restoration of normal vesicle numbers. Results of *Zimmermann* and *Whittaker* (1974 b)

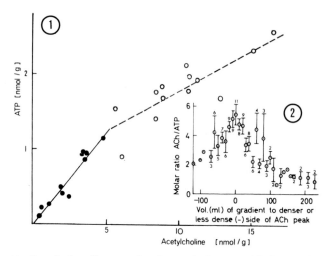

Fig. 12. (1) Correlation diagram showing vesicular acetylcholine and ATP concentrations in the peak vesicular acetylcholine fraction in control (open circles) and stimulated (filled circles) organs. Note the close correlation between vesicular acetylcholine and ATP levels in stimulated organs. (2) Acetylcholine: ATP ratios plotted across the vesicular acetylcholine peak. The fractions are identified as the volume by which they lie to the rotor centre or periphery from the peak acetylcholine tube. Bars are ± S.E.M. of mean values given by the points; figures give the number of experiments meaned

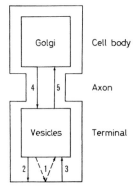

Fig. 13. Processes which may influence the steady state population of vesicles in the nerve terminal: (1) transient exocytosis of vesicles leading to progressive loss of structure and ability to store transmitter; (2) complete (full) exocytosis, leading to loss of vesicles by incorporation of vesicle membrane with the external pre-synaptic membrane; (3) endocytosis leading to reformation of vesicles difficult to distinguish (except by incorporation of endocytotic markers such as horseradish-peroxidase and ferritin) from normal synaptic vesicles but with unproven ability to take up transmitter; (4) centrifugal axonal flow of vesicles recently synthesized in the cell body; (5) centripetal axonal flow, possibly conveying salvaged vesicle membrane to cell body for reprocessing

content of the terminals. Maximum responses to single shocks are associated with a vesicle population with a high acetylcholine: ATP ratio: such vesicles are the first to disappear on stimulation and reappear on resting.

Axonal Flow in the Electromotor Nerves

Clearly, the total vesicle population at the nerve terminals is in a dynamic state of flux; Fig. 13 expresses diagrammatically some of the processes discussed in recent literature (*e.g. Bloom, Iversen* and *Schmitt*, 1970) which may be involved. Stimulation may well perturb the steady state vesicle population by promoting exocytosis and also, directly or indirectly, by stimulating axonal flow. In order to obtain more information about the possible contribution of axonal flow to the dynamics of the vesicle population we have made preliminary studies of the accumulation of acetylcholine and vesicular protein above a ligature placed around the electromotor nerve trunks. Axonal flow of acetylcholine, choline acetyltransferase and cholinesterase are well-known phenomena in mammalian cholinergic axons (see *Kása, Mann, Karcsu, Tóth* and *Jordan*, 1973; *Saunders, Dziegielewska, Häggendal* and *Dahlström*, 1973; and references quoted therein).

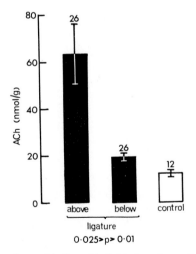

Fig. 14. Accumulation of acetylcholine (black blocks) above (left) and below (right) a ligature, compared with acetylcholine levels in normal nerve (white blocks). Values are means ± S.E.M. (shown by the vertical bars) of the number of experiments shown above each block. The ligatures were in position for 3 days and assays were performed with the leech microassay on extracts of 25 mg of nerve trunk

Acetylcholine

Fig. 14 shows that acetylcholine accumulates above, and to a limited extent, below a ligature in these nerves. The raised levels of acetylcholine below the ligature may represent leakage of acetylcholine from the region of high acetylcholine above the ligature, but could conceivably represent reversed (centripetal) axonal flow.

The acetylcholine values given in Fig. 14 are probably considerably higher than the true values. It has been observed that about only 50 % of the apparent acetylcholine in the nerve extracts as assayed on the dorsal muscle of the leech is recoverable by gas chromatography (estimations kindly performed by Prof. B. Holmstedt) and destroyed by acetylcholinesterase. However, when allowance is made for the so far unidentified interfering substance, the basic finding of acetylcholine accumulation above the ligature remains unaffected.

An experiment in which stimulation was applied before the ligature did not significantly increase axonal flow (Table 2).

Table 2. *Acetylcholine Accumulation above a Ligature in Stimulated and Unstimulated Electromotor Nerves*

| | Acetylcholine content (nmol/g wet wt.) | | | |
| | No stimulation | | Stimulation | |
Nerve	Above	Below	Above	Below
Left I	43	6	118	56
Left III	130	31	372	14
Left IV	—	26	171	52
Right I	38	30	34	16
Right III	36	21	53	—
Right IV	389	23	53	31
Mean	127	23	133	34

Sign test shows a highly significant difference ($p = 0.002$) for the 10 out of 10 positive differences between the regions above and below the ligature. Stimulation, however, exerted no significant effect. Results of *Zimmermann* and *Whittaker* (1973 a).

Double ligature experiments will be needed to determine the proportion of axonal acetylcholine that is subject to axonal flow and the rate of this flow.

Vesicle Membrane Proteins

Vesicle membrane proteins are antigenic (*Widlund et al.*, 1974; *Ulmar* and *Whittaker*, 1974 a); using the indirect immunohisto-fluorescence method (Fig. 15) it has been possible to observe the accumulation of antigenic vesicle protein above a ligature (*Ulmar* and *Whittaker*, 1974 b) (Fig. 16). The technique has also enabled cholinergic nerve fibres in the electric organ, electromotor nucleus and the heart of *Torpedo* to be visualized; unfortunately the anti-serum did not cross react with cholinergic vesicles from other species so could not be used as the basis of a general method for visualizing cholinergic nerves.

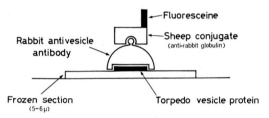

Fig. 15. Summary of the procedure involved in immunohistofluorescence studies by the indirect method. Rabbit antivesicle antibody becomes bound to antigen wherever this may occur in a section. After washing away non-specifically absorbed antibody, the location of the remaining antibody is visualized by its ability to bind sheep antirabbit globulin labelled with fluoresceine

Morphological Effects

Fig. 17 shows the morphological changes which occurred in the axoplasm in the region immediately above (Fig. 17-2), and below (Fig. 17-3) relative to normal cytoplasm (Fig. 17-1). As has also been observed by others, the affect of ligaturing is to cause the appearance of large numbers of vesicles of very varying size on *both* sides of the ligature; dense particles (possible small lysosomes) also make their appearance. Due to the profusion of vesicular structures present in the cytoplasm of the ligatured nerve it is not possible to show an accumulation of *synaptic* vesicles morphologically above the ligature, but there certainly are many vesicles of the right size range present. The biochemical results demonstrate that axonal flow of both acetylcholine and vesicle membrane protein occur, but more work will be needed to determine the quantitative contribution of axonal flow to the maintenance of the normal vesicle population at the terminals. Preliminary measurements of the non-cytoplasmic acetyl-

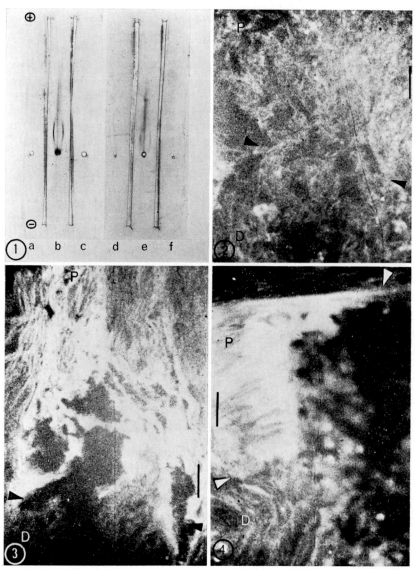

Fig. 16. (1) Precipitin patterns obtained by immunoelectrophoresis in agar with *Torpedo* cholinergic vesicle antiserum. Two vesicle fractions from different antigen preparations were placed in wells b and e: note one strong and 1—2 very faint lines. No precipitation is seen with soluble cytoplasmic proteins (well a, fraction SP), fragments of postsynaptic membranes (well c, fraction MP), chromaffin granules (well d) or guinea pig cerebral cortical synaptic vesicles (well f). The troughs contained immune serum. (2) Control longitudinal section showing no accumulation of fluorescence in a ligated electromotor nerve trunk after incubation with pre-immune serum. (3, 4) Increased specific immunofluorescence proximal to the ligation (shown by arrows). In (4) the proximal part of the axon is distorted to the left of the micrograph. The bar is 100 μm

choline: ATP ratio in the axon suggest that the "acetylcholine-rich" vesicles are formed in the terminal, perhaps when newly-arrived vesicles come into contact with a relatively high concentration of cytoplasmic acetylcholine.

Conclusions

Many questions remain to be answered concerning the detailed mechanisms involved in transmitter storage and release. The findings with a purely cholinergic system, the electromotor nerves of the *Torpedo*, indicate that cholinergic neurones may have more points of resemblance to adrenergic neurones than was previously thought. The vesicles in both types of neurone appear to be heterogeneous (*de Potter et al.*, 1972); both seem to be transported by axonal flow. Both contain acidic core proteins [though these are not identical, either in amino acid composition (*Whittaker*, 1974) or immuno-chemically (*Ulmar* and *Whittaker*, 1974 a)] in combination with ATP as well as transmitter; the function of the ATP protein complex appears to be the provision of negative charges to neutralize the biogenic amine but other roles (energy source, trophic factor) for the ATP are not excluded. Variations in transmitter: ATP ratios appear to be one way in which vesicle heterogeneity expresses itself in both systems: another may be the rate of turnover of transmitter.

The actual mechanism of unloading and refilling of the vesicles (if the latter occurs) is still obscure. An ion-exchange reaction during transient exocytosis in which Ca^{2+} or Na^+ from the extracellular fluid displace the acetylcholine ion is one possibility; suitable carriers in the external membrane of the vesicle might permit spontaneous reloading in regions of high cytoplasmic acetylcholine by efflux of such inorganic ions. But are vesicles reused, and if so, how many times; and what causes complete exocytosis of the vesicle membrane during repetitive stimulation? Is this fate limited to exhausted vesicles that have already undergone several cycles of transient exocytosis or is it a direct response to strong stimulation? Are the "empty" vesicles reformed after stimulation capable of acquiring transmitter anew in the periphery or do they represent merely a membrane salvage operation, a resorption of excess external membrane essential to the restoration of the prestimulation configuration of the synapse? If the latter how is the salvage operation completed: are the empty vesicles returned by centripetal axonal flow to the cell body for eventual fusion with the Golgi membranes? The techniques now exist for a direct experimental attack on some of these problems.

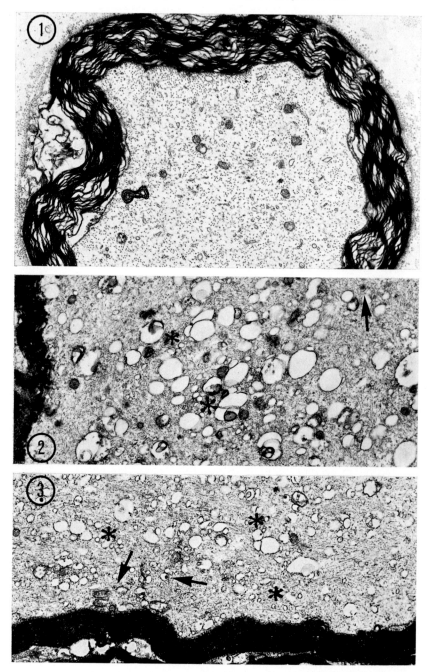

References

Bloom, E. F., L. L. Iversen, and *F. O. Schmitt:* Macromolecules in synaptic function. Neurosci. Res. Bull. *8* (1970).

Changeux, J.-P.: Études sur le mécanisme moléculaire de la résponse d'une membrane excitable aux agents cholinergiques. In: Le Système Cholinergique en Anesthésiologie et en Réanimation (*Nahas G.-G., J.-C. Salamagne, P. Viars,* and *G. Vourc'h,* eds.), pp. 99—112. Paris: Librairie Arnette. 1972.

Da Prada, M., and *A. Pletscher:* Isolated 5-hydroxytryptamine organelles of rabbit blood platelets: physiological properties and drug-induced changes. Brit. J. Pharmacol. *4,* 591—597 (1968).

de Potter, W. P., I. W. Chubb, and *A. F. de Schaepdryver:* Pharmacological aspects of peripheral noradrenergic transmission. Arch. int. Pharmacodyn. Ther. *196* (suppl.), 258—287 (1972).

de Potter, W. P., D. P. Smith, and *A. F. de Schaepdryver:* Subcellular fractionation of splenic nerve: ATP, chromogranin A and dopamine β-hydroxylase in noradrenergic vesicles. Tissue and Cell 2, 529—546 (1970).

Dowdall, M. J., A. F. Boyne, and *V. P. Whittaker:* Adenosine triphosphate: a constituent of cholinergic synaptic vesicles. Biochem. J. *140,* 1—12 (1974).

Fritsch, G.: Die elektrischen Fische. Zweite Abtheilung: Die Torpedineen. Leipzig: von Veit & Co. 1890.

Geffen, L. B., and *B. G. Livett:* Synaptic vesicles in sympathetic neurones. Physiol. Rev. *51,* 98—157 (1971).

Heilbronn, E., and *H. Petterson:* Acetylcholine and related enzymes in normal and ligated cholinergic nerves from *Torpedo marmorata.* Acta Physiol. Scand. *88,* 590 (1973).

Hillarp, N.-Å.: Adenosine phosphates and inorganic phosphate in the adrenaline and noradrenaline containing granules of the adrenal medulla. Acta Physiol. Scand. *42,* 321—332 (1958).

Karlsson, E., E. Heilbronn, and *L. Widlund:* Isolation of the nicotinic acetylcholine receptor by biospecific chromatography on insolubilized *Naja naja* neurotoxin. FEBS Letters *28,* 107 (1972).

Kása, P., S. P. Mann, S. Karcsu, L. Tóth, and *S. Jordan:* Transport of choline acetyltransferase and acetylcholinesterase in the rat sciatic nerve: a biochemical and electron histochemical study. J. Neurochem. *21,* 431 to 436 (1973).

Miledi, R., P. Molinoff, and *L. T. Potter:* Isolation of the cholinergic receptor protein of *Torpedo* electric tissue. Nature *229,* 554 (1971).

Saunders, N. R., K. Dziegielewska, C. J. Häggendal, and *A. B. Dahlström:*

Fig. 17. Morphological effects of ligature in an electromotor nerve of *Torpedo.* Note vesicular structures in the axoplasm some of which (asterisk) are in the size range of synaptic vesicles both (2) above and (3) below a ligature as compared with (1) normal axoplasm. Small dense bodies (lysosomes?) also make their appearance (arrows). ×10,000

Slow accumulation of choline acetyltransferase in crushed sciatic nerves of the rat. J. Neurobiol. *4*, 95—103 (1973).

Schmidt, J., and *M. A. Raftery:* Use of affinity chromatography for acetylcholine receptor purification. Biochem. Biophys. Res. Comm. *49*, 572 (1972).

Schümann, H. J.: Über den Noradrenalin- und ATP-Gehalt sympathischer Nerven. Archiv. exp. Path. u. Pharmacol. *233*, 296—300 (1958).

Smith, A. D.: Biochemistry of adrenal chromaffin granules. In: The Interaction of Drugs and Subcellular Components of Animal Cells (*Campbell, P. N.*, ed.), pp. 239—292. Boston: Little Brown & Co. 1968.

Soifer, D., and *V. P. Whittaker:* Morphology of subcellular fractions derived from the electric organ of *Torpedo*. Biochem. J. *128*, 845—846 (1972).

Ulmar, G., and *V. P. Whittaker:* Immunological approach to the characterization of cholinergic vesicular protein. J. Neurochem. *22*, 452—455 (1974 a).

Ulmar, G., and *V. P. Whittaker:* Immunohistochemical localization and immunoelectrophoresis of cholinergic synaptic vesicle protein constituents from the *Torpedo*. Brain Res. *71*, 155—159 (1974 b).

Whittaker, V. P.: Origin and function of synaptic vesicles. Ann. N.Y. Acad. Sci. *183*, 21—32 (1971).

Whittaker, V. P.: The storage of acetylcholine in presynaptic nerve terminals. In: The Scientific Basis of Medicine Annual Reviews 1973 (*Gilliland, I.*, and *M. Peden*, eds.), pp. 17—31. London: Athlone Press. 1973 a.

Whittaker, V. P.: The structural and chemical properties of synaptic vesicles. In: Proteins of the Nervous System (*Schneider, D.*, ed.), pp. 155 to 169. New York: Raven Press. 1973 b.

Whittaker, V. P.: Molecular organization of the cholinergic vesicle. Adv. Cytopharmacol. *2*, 311—317 (1974).

Whittaker, V. P., and *Dowdall, M. J.:* Constituents of cholinergic vesicles. In: La Transmission cholinergique de l'Excitation (*Fardeau, M.*, *M. Israël*, and *R. Manaranche*, eds.), pp. 101—117. Paris: Editions INSERM. 1973.

Whittaker, V. P., *M. J. Dowdall*, and *A. F. Boyne:* The storage and release of acetylcholine by cholinergic nerve terminals: recent results with non-mammalian preparations. Biochem. Soc. Symp. *36*, 49—68 (1972 a).

Whittaker, V. P., *W. B. Essman*, and *G. H. C. Dowe:* The isolation of pure cholinergic synaptic vesicles from the electric organs of elasmobranch fish of the family Torpedinidae. Biochem. J. *128*, 833—846 (1972 b).

Whittaker, V. P., *M. J. Dowdall*, *G. H. C. Dowe*, *R. M. Facino*, and *I. Scotto:* Proteins of cholinergic synaptic vesicles from the electric organ of *Torpedo:* characterization of a low molecular weight protein. Brain Res. *75*, 115—131 (1974).

Widlund, L., *K. A. Karlsson*, *A. Winter*, and *E. Heilbronn:* Immunochemical studies on cholinergic synaptic vesicles. J. Neurochem. *22*, 451 to 456 (1974).

Zimmermann, H., and *V. P. Whittaker:* Evidence for axonal flow of acetylcholine (ACh) in cholinergic nerves. Abstr. 4th int. Meet. int. Soc. Neurochem. Tokyo, p. 245 (abstr. no. 223) (1973 a).

Zimmermann, H., and *V. P. Whittaker:* The effect of stimulation on the composition and yield of cholinergic synaptic vesicles. Abstr. 4th int. Meet. int. Soc. Neurochem. Tokyo, p. 321 (abstr. no. 343) (1973 b).

Zimmermann, H., and *V. P. Whittaker:* Effect of electrical stimulation on the yield and composition of synaptic vesicles from the cholinergic synapses of the electric organ of *Torpedo:* a combined biochemical, electrophysiological and morphological study. J. Neurochem. *22,* 435 to 451 (1974 a).

Zimmermann, H., and *V. P. Whittaker:* Different recovery rates of the electrophysiological, biochemical and morphological parameters in the cholinergic synapses of the *Torpedo* electric organ after stimulation. J. Neurochem. *22,* 1109—1114 (1974 b).

Authors' address: Abteilung für Neurochemie, Max-Planck-Institut für biophysikalische Chemie, Postfach 968, D-3400 Göttingen, Federal Republic of Germany.

Discussion

Sotelo: Is the heterogeneity of synaptic vesicles that you have demonstrated by comparing control with stimulated axon terminals in Torpedo electric organ, related to the two pools of acetylcholine—the labile and the stable one—reported for this material; and if so, have you found the two vesicular populations in non-stimulated animals?

Whittaker: I presume that by "labile" and "stable" acetylcholine you mean those fractions of the total acetylcholine of the tissue that are, respectively, destroyed by acetylcholinesterase on homogenization of the tissue and are unaffected by acetylcholinesterase. Since our vesicles are isolated from a cytoplasmic extract which is very rich in acetylcholinesterase (we do not add esterase inhibitors), our vesicular acetylcholine is, by definition, "stable"; it is also, as we have seen, heterogeneous.

The origin of the "labile" pool is hard to determine. Since in the intact organ, terminal cytoplasm is physically protected from the action of acetylcholinesterase by the external presynaptic terminal membrane, and since choline acetyltransferase is a cytoplasmic enzyme, one predicts the existence of a cytoplasmic pool of acetylcholine which would be destroyed (hence be "labile") on homogenization. However, homogenization might also release some vesicular acetylcholine, and this would also be estimated with the "labile" pool. The proportions of cytoplasmic and vesicular acetylcholine contributing to the "labile" pool are impossible to determine at present and may well vary according to the homogenization conditions used.

Thoenen: You have provided evidence that in consequence of electrical nerve stimulation the membranes of the vesicles are incorporated into the neuronal membrane of the cholinergic nerve terminals of the *Torpedo*. Is there any evidence that vesicles are reformed from the neuronal membrane and if yes are these vesicles able to release transmitter again?

Whittaker: Within twenty-four hours after stimulating to exhaustion the morphology of the nerve terminals returned to normal, the invaginations of the presynaptic membrane disappeared and the cytoplasm of the nerve terminals was again filled with vesicles (Fig. 11). However, it takes about three days for the vesicular acetylcholine and ATP to recover and during this period when the biochemical and morphological recovery is out of phase the organ is abnormally fatiguable. That is, the first impulse gives a normal response but on repetative stimulation the response declines much more rapidly than it would do under normal conditions.

Journal of Neural Transmission, Suppl. XII, 61—74 (1975)
© by Springer-Verlag 1975

Quantum Hypothesis of Synaptic Transmission

A. Wernig

Pharmakologisches Institut, Universität Innsbruck, Austria[*]

The Binomial Nature of Transmitter Release

The quantum hypothesis of transmitter release states that transmitter is released from presynaptic nerve terminals solely in transmitter packages (quanta). Experimental evidence for this is to be found in the occurrence of spontaneous depolarizations of the postsynaptic membrane (spont. e.p.p.s) with normally distributed amplitudes (unit potentials) (*Fatt* and *Katz*, 1952). Furthermore one can show that the postsynaptic membrane depolarization after the stimulation of the presynaptic nerve (evoked e.p.p.) is built up from single unit depolarizations (*del Castillo* and *Katz*, 1954; *Boyd* and *Martin*, 1956). Convincing direct evidence for quantal release comes from experiments at low temperature where the release of transmitter is sufficiently dispersed in time, so that by recording synaptic activity with extracellular glass microelectrodes it is possible to count individual quanta as they are released following a nerve impulse (*Katz* and *Miledi*, 1965 b). In a large number of nerve impulses under constant conditions the number of quanta released per impulse fluctuates from trial to trial to give a certain release pattern (*i.e.* observed numbers of trials which released 0, 1, 2, 3 a.s.o. quanta in the series) so that transmitter release might be considered a statistical process. The model proposed by *del Castillo* and *Katz* (1954) to describe the release process assumed the presence of a number (n) of units within the nerve terminal, each having a certain probability of release in connection with an oncoming nerve impulse. If the average of such probabilities is p, the average number of

* Present address: Max-Planck-Institut für Psychiatrie, Abteilung Neurophysiologie, Kraepelinstraße 2, D-8000 München 40, Federal Republic of Germany.

quanta released per impulse (m) during a series of trials is given by

$$m = n \times p . \tag{1}$$

Assuming binomial statistics the expected numbers of trials (n_x) which release x quanta (expected release pattern) should be predicted by

$$n_x = N \times \frac{n!}{(n - x)!\, x!}\, p^x\, (1 - p)^{n-x} , \tag{2}$$

where N is the total number of trials in the series. If p is low ($p \to 0$), a binomial distribution can be approximated by Poisson statistics and

$$n_x = N \times \frac{e^{-m} \times m^x}{x!} . \tag{3}$$

It is obvious from Eqn. 3 that for calculating the expected release pattern in Poisson statistics only m is required from experimental observation, in contrast to binomial statistics, which require the determination of both, p and n. Del Castillo and Katz (1954) used Poisson statistics because transmitter release had been markedly reduced in their experiments by adding Mg (and decreasing the Ca concentration). They assumed that Mg decreased p (from Eqn. 1), so that $p \to 0$. m was estimated from the mean amplitude of the evoked e.p.p.'s in the series ($\bar{v}$), and the mean amplitude of the unit potential as measured from the spontaneous e.p.p.s (v_1),

$$m = \bar{v}/\bar{v}\,_1 . \tag{4}$$

The expected release pattern was thus calculated and comparison with the observed amplitude distribution resulted in a good agreement in many cases (*del Castillo* and *Katz*, 1954; *Boyd* and *Martin*, 1956). This indicates, that the evoked response of the postsynaptic membrane was indeed built up of single units which are identical in amplitude with the spontaneously occuring depolarizations. Such observations have since been made in a wide variety of vertebrate and invertebrate synapses (for reviews see *Martin*, 1966; *Kuno*, 1971).

However, there have also been reports on deviations from Poisson statistics (*Atwood* and *Parnas*, 1968; *Atwood* and *Johnston*, 1968; *Auerbach* and *Bennett*, 1969; *Blackman*, *Ginsberg* and *Ray*, 1963; *Blackman* and *Purves*, 1969; *Bittner* and *Harrison*, 1970; *Ishii*, *Matsumura* and *Furukawa*, 1971; *Kuno* and *Weakly*, 1972). The theory originally proposed by *del Castillo* and *Katz* (1954) was essential binomial and Poisson statistics were thought to be applicable only in the specific situation of a low p; therefore deviations from Poisson statistics do not necessarily violate the quantum hypothesis.

Besides measuring the mean number of quanta released per trial (m), the variance (var) of the quantal fluctuation can be estimated from experimental observation. In a binomial distribution var is also given by

$$var = n \times p \, (1 - p) \, . \qquad (5)$$

Combining Eqns. 1 and 5, p and n can be calculated from

$$p = 1 - \frac{var}{m} \qquad (6)$$

and

$$n = m/p \, . \qquad (7)$$

Such estimations were first performed in experiments on crayfish neuromuscular junction in low temperature, where transmitter release was dispersed in time so that the number of quanta released per trial could be counted (*Johnson* and *Wernig*, 1971). Using the estimated values for p and n, the release pattern was calculated from Eqn. 2. The results showed that for all values of m and p the observed release pattern was approximated closely by the corresponding binomial distribution. However, Poisson predictions differed significantly from the observed quantal distribution for higher values of p (*Johnson* and *Wernig*, 1971; *Zucker*, 1973).

The evaluation of p and n in experiments with different Ca and Mg concentrations gave direct evidence for the action of Mg and Ca on the statistical release parameter p. The decrease in transmitter release in high Mg concentration occured in parallel with the decrease in p and the increased transmitter output in high Ca went parallel with an increase in p (*Wernig*, 1972 b). This provides direct evidence for the validity of the original assumption by *del Castillo* and *Katz* (1954), that Mg reduced p and made Poisson statistics applicable in their experiments.

The application of binomial statistics instead of Poisson statistics may have important consequences, as will be outlined in the next chapter. Therefore the question seems important whether binomial release can also be demonstrated in other synapses than the crayfish neuromuscular junction.

Assuming that the release is a binomial process, *Christensen* and *Martin* (1970) estimated p from the statistical analysis of intracellular recordings from neuromuscular junctions in the rat diaphragm. They measured the means and variances of e.p.p. amplitudes at two Ca concentrations, assuming that Ca only changes p. This measure of p is indirect, but their results suggest that transmitter release at the diaphragm might be a binomial process. *Kuno* and *Weakly* (1972) analyzed the inhibitory synaptic potential (i.p.s.p.) in spinal motoneurons of the cat and found marked deviations from Poisson statistics. The amplitude distribution of evoked i.p.s.p.'s could be approximated accurately by assuming binomial statistics, thus providing evidence for the quantal nature of the inhibitory synaptic potential.

Statistical analysis of transmitter release at unblocked (normal Ca and Mg concentration of the bathing solution) frog neuromuscular junction is difficult since the endplate potential evoked by a single nerve impulse normally reaches the threshold for the action potential and its amplitude is therefore not measurable. If, on the other hand, synaptic transmission is blocked by curare, the unit potential (spon. e.p.p.) can not be measured. To avoid these difficulties, *Katz* and *Miledi* (1965 a, b) recorded from frog neuromuscular synapse synaptic activity evoked by nerve impulses in Ca-free solutions. Extracellular glass microelectrodes were filled with $CaCl_2$ and thus were used both for recording and providing a local supply of Ca. In this way transmitter release was confined to a limited area under the recording electrode and the observed mean quantum content was reduced to 0.41—1.36 quanta per impulse. Experiments were performed at low temperature (2—17.5 °C) so that release was sufficiently dispersed in time for released quanta to be distinguished individually. In these experiments Poisson statistics were sufficiently accurate except for one experiment in which dispersion of the postsynaptic response might not have been sufficient because of a relatively high temperature (17.5 °C) (*Katz and Miledi*, 1965 a; Table 1). Recalculation of this experiment based on binomial statistics provides a better approximation to the observed release pattern. (This can be expected however, if the number of high quantal releases was underestimated and the number of single unit releases overestimated.)

Katz and *Miledi* (1967 a, b, c) described another method for limiting the transmitter release to a small area of the frog neuro-muscular synapse. When action potentials are abolished by tetrodo-toxin, transmitter release can still be induced focally by passing de-polarizing current through a glass microelectrode (with a tip diameter of a few microns) located over a nerve terminal branch. Synaptic activity can then be monitored at normal temperature with an intracellular electrode. Such experiments can be performed to estimate the statistical release parameters n and p from intracellular recordings

(*Wernig*, 1973). In order to estimate the variance (*var*) and *m* of the quantal fluctuation from the amplitude distribution of the evoked response (after correction for nonlinear summation, *Martin*, 1955) the e.p.p. histogram was divided into groups with the mean amplitude of the spont. e.p.p.'s ($\bar{v}_1$) as the mean of the first group (single unit releases), two times ' $_1$ was used as the mean of the second group (double unit releases) a.s.o. The numbers of observations within groups were counted and *m* and *var* calculated from these numbers (*m* = total number of quanta released/N). *p* and *n* were then calculated from Eqns. 1 and 5. Another way of estimating *p* and *n* can be used in certain experiments by using the binomial prediction for the number of transmission failures (*Wernig*, 1973). From Eqn. 2

$$n_0 = N \times (1 - p)^n , \qquad (8)$$

where n_0 is the expected number of failures. By taking the observed number of failures as n_0, Eqn. 8 can be used instead of *var* (and Eqn. 5). Since m can also be obtained directly from amplitude measurements (Eqn. 4 after correction for nonlinear summation), there is no need in this case for "grouping" the amplitude histogram. These two independent measures of the statistical release parameters gave similar results and indicated deviations from Poisson statistics (*Wernig*, 1973). A major disadvantage of this method of focal stimulation is the lack of a measure of the induced presynaptic depolarization which, however, is an important factor regulating the secretion of the transmitter substances (see *Katz* and *Miledi*, 1967 a). Therefore no direct comparison can be made with transmitter release evoked by the physiological depolarization of action potentials.

In order to estimate statistical release parameters of a whole frog neuromuscular synapse from intracellular recordings, the following considerations might be helpful. Since the coefficient of variation (CV) is independent of the unit of measurement, the coefficient of variation of the amplitude distribution should be identical with the coefficient of variation of the quantum content distribution (*Martin*, 1966). Allowance can be made for the fact that the unit potentials themselves are not of identical size. This can be done by a factor $\sqrt{1 + CV^2}$, where CV is the coefficient of variation of the unit potential amplitude. In practice this factor may be ignored (see *Martin*, 1966). Its omission will lead to a slight underestimate of *p*.

$$CV^2 = \frac{\sigma^2}{\bar{v}^2} = \frac{var}{m^2} ,$$

where σ^2 is the variance of the amplitude distribution. Using m from Eqn. 4,

$$var = \frac{\sigma^2}{\overline{v}_1^2} \tag{9}$$

and from Eqn. 6 it follows, that

$$p = 1 - \frac{\sigma^2}{\overline{v}_1 \times \overline{v}} . \tag{10}$$

Such calculations were performed in the experiment listed below. Intracellular recordings were made from a neuromuscular junction in isolated frog cutaneous pectoris muscle. Transmitter release was evoked by stimulating the dissected nerve trunk with conventional Pt-electrodes. Transmitter release was sufficiently reduced by adding Mg to avoid twitching of almost all muscle fibers. Resting potential was measured to be 100 mV and stayed constant throughout the experiment. 275 shocks were applied to the nerve at continuous stimulation at 1/sec and e.p.p.'s were recorded on moving film from a Tektronix 565 oscilloscope and measured from the film records. Measurements of e.p.p.'s were individually corrected for nonlinear summation (*Martin*, 1955). No obvious drift in release was found in consecutive groups of 25 trials. 126 spontaneous e.p.p.'s were measured and were found to be normally distributed with a mean amplitude of 0.375 mV.

Table 1. *Release Parameters from Experiment on Mg Blocked Frog Neuromuscular Junction. All Values Were Calculated after e.p.p.s Had Indivudually Been Corrected for Nonlinear Summation (Martin, 1955), p and n Were Calculated from Eqns. 10 and 7. $\overline{v}$ and $\overline{v}_1$ Show the Corrected Values for the Mean Amplitude of the Evoked e.p.p.s and the Mean Amplitude of the Spontaneous e.p.p.s Respectively. $m = \overline{v}/\overline{v}_1$*

Resting potential (mV)	$\overline{v}_1$ (mV)	$\overline{v}$ (mV)	m	p	n	N
100	0.375	16.31	43.48	0.34	127	275

In this experiment p was calculated to be 0.34. If the reduction of transmitter release by Mg also in frog goes parallel with the decrease in p, as was found in crayfish neuromuscular junction (*Wernig*, 1972 b), even higher values for p than found in the present experiment have to be expected for synapses where the release is not blocked by Mg.

Such results would indicate, that at the frog neuromuscular junction transmitter release is governed by binomial statistics.

Consequences of Non-Poisson Release

It has become widely accepted to estimate m (the mean number of quanta released per trial in a series) in a number of different ways. The best way to obtain the number of quanta released by a single nerve impulse is to observe individually the postsynaptic effect of each quantum. This is possible at low temperature where the delay and the time course of transmitter release is prolonged, so that the number of quanta released can be counted directly (*Katz* and *Miledi*, 1965 a, b). It is understandable, however, that this technique has to be limited to experiments with low quantal release (< 6—8 quanta). m can be estimated as the total number of quanta released divided by the total number of trials. As pointed out above, the amplitude of the spont. e.p.p.'s from intracellular recordings is not uniform, but distributed according to a Gaussian curve. Therefore when there is a large number of quanta in the evoked synaptic potential, it becomes impossible to tell the exact number involved. When a large number of e.p.p.'s is evoked, however, m can be calculated from the mean amplitude of the evoked e.p.p.'s ($\bar{v}$) divided by the mean amplitude of the spont. e.p.p.'s ($\bar{v}_1$) (Eqn. 4). Of course, nonlinear summation must be taken into account. After *del Castillo* and *Katz* (1954) showed that fluctuations in release obeyed statistics based on Poisson's Law, and such calculations had been found to be applicable in many synapses, it became convenient to utilize features of Poisson statistics for estimating m. One such statistical estimation is based on the observed number of transmission failures and is derived from Eqn. 3:

$$m_0 = \ln N/n_0 .$$

Another frequently used estimation of m is based on the co-efficient of variation (CV). In Poisson statistics the variance is equal to the mean, so

$$m_2 = 1/CV^2 . \qquad (11)$$

Unless values of p are small, these estimates of m which are strictly based on the assumption of a Poisson distribution may be considerably in error (*Johnson* and *Wernig*, 1971). The corresponding binomial expressions are

$$m_0 Bin = \frac{-p}{\ln (1-p)} \ln N/n_0 \qquad (12)$$

and

$$m_2 Bin = \frac{(1-p)}{CV^2} . \qquad (13)$$

It becomes obvious from these equations, that Poisson calculations overestimate m by 4 to 39 % and 11 to 100 % respectively for values of p ranging from 0.1 to 0.5. To illustrate this, several estimations of m are compared in Table 2. Data are recalculated from experiments on crayfish neuromuscular junction in low temperature, where p was calculated in the above described way (*Johnson* and *Wernig*, 1971, Table 2).

Table 2. *Different Estimates of m for 4 Experiments. Values for m Are Recalculated from Experiments on Crayfish Neuromuscular Junction (Johnson and Wernig, 1971, Table 2). p Had Been Estimated Using Eqn. 6. m Is Calculated from Total Number of Quanta Released in a Series Divided by N, m_0 and m_2 Are Based on Poisson Assumptions, m_0Bin and m_2Bin Are the Corresponding Binomial Estimations (Eqns. 12 and 13)*

Experiment	m	m_0	m_2	m_0Bin	m_2Bin	p
III	0.36	0.37	0.38	0.35	0.36	0.08
IV	0.70	0.74	0.79	0.70	0.70	0.11
V	0.88	0.96	1.11	0.86	0.88	0.21
VI	1.36	1.91	2.69	1.39	1.36	0.49

As expected, m_2 and m_0 increasingly overestimated m (compared with the total number of quanta released/N) as p increased. The difference was abolished by using Eqns. 12 and 13 which include the calculated values for p (m_0Bin and m_2Bin in Table 2). It has also become convenient to calculate rather than to measure the mean size of the unit potential (q) in certain situations (*Elmqvist* and *Quastel*, 1965 a, b). This might be required when the amplitude of the spont. e.p.p.'s is too small to be clearly distinguishable from baseline noise, e.g. when postsynaptic sensitivity is decreased by curare. Using Poisson statistics, q can be calculated as

$$q = \frac{\sigma^2}{\overline{v}} \qquad \text{(from Eqns. 4 and 11).}$$

The corresponding binomial term is

$$q = \frac{\sigma^2}{\overline{v}\,(1 - p)} \,.$$

In fact, *Elmqvist* and *Quastel* (1965 b) observed in experiments on the human neuromuscular synapse that q increased with increasing frequency of stimulation from 0.2 to 160 stimuli per sec (*Elmqvist* and *Quastel*, 1965 b, Fig. 3 B). Since other reasons for this change in q seem unlikely, this result can be interpreted as a change in p due to the increased frequency of stimulation.

The conclusion from such results is that statistical estimates of m and q must include p unless existence of a Poisson distribution is clearly evident. This seems especially desirable in experiments in which induced changes in transmitter release might go parallel with changes in p (changing Ca and Mg concentration, frequency of stimulation).

Morphological and Functional Considerations and the Statistical Release Parameters

If it is accepted that transmitter release is governed by binomial statistics, it becomes important to know, whether any functional or morphological meaning can be attached to n and p. Quite a few transmitter release theories, some based on binomial assumptions, have been developed. It would lead too far to discuss these here, especially since excellent reviews have been written (*Ginsborg*, 1970; *Hubbard*, 1970; *Kuno*, 1971). A few remarks might be allowed considering the possible meaning of p and n in the release process. As *Zucker* (1973) has recently pointed out it is important to stress that n and p (as defined here) must not be confused with nominally identical parameters of transmitter release reported in the literature. In these cases n is thought to be a certain number of transmitter quanta in the presynaptic terminal which can be estimated from the initial decline of e.p.p.'s following the first e.p.p. in continuous stimulation (*e.g. Elmqvist* and *Quastel*, 1965 b).

If the values for n and p are governed by any functional or morphological features, one should expect that changes in the amount of transmitter release induced by physiological or experimental means should be reflected in the statistical estimates of p and n. It is well known that transmitter release is greatly dependent on external Ca and Mg concentrations. Increasing the Ca concentration in crayfish neuromuscular junction leads to an increase in the number of quanta released per impulse (m) (*Bracho* and *Orkand*, 1970; *Wernig*, 1972 b). Statistical analysis of transmitter release suggested that in all experiments increases in m in higher Ca concentrations were due to an increase in p (*Wernig*, 1972 b). Similarly the depressing action of Mg went parallel with the decrease in p. Facilitated release of transmitter in crayfish is also due to an increase in m (*Dudel* and *Kuffler*, 1961; *Dudel*, 1965; *Bittner*, 1968; *Bittner* and *Harrison*, 1970; *Bittner* and *Kennedy*, 1970; *Atwood* and *Bittner*, 1971; *Wernig*, 1972 a; *Zucker*, 1973). Results from continuous stimulation at different frequencies (*Wernig*, 1972 a; *Zucker*, 1973) and from double shock experiments (*Zucker*, 1973) indicate that facilitation in crayfish is accompanied by an increase in p. Such results show that the statistical release parameters p and n indeed unanimously follow changes in experimental conditions. One might think of n as a number of quanta stored in the nerve terminal. The values of n calculated from crayfish neuromuscular junction were between 2 and about 10 (*Johnson* and *Wernig*, 1971; *Wernig*, 1972 a, b; *Zucker*, 1973) and therefore do not favour such an idea. Similarly n was estimated to be less than 4 at inhibitory synapses on spinal motoneurons of the cat (*Kuno* and *Weakly*, 1972). In frog neuromuscular junction values for n can be calculated from the above described experiments by *Katz* and *Miledi* (1965 a, b), showing values for n between 3 and 13. The actual length of nerve terminal involved in such experiments is not known but is probably of the order of 15—20 microns (*del Castillo* and *Katz*, 1956). It is interesting, therefore, to compare these values with the estimated n of more than 100 from the experiment on frog neuromuscular junction described above, where transmitter release was monitored from the whole length of a synaptic contact. First the comparison indicates that n is dependent on the physical dimension of the releasing area. When morphological correlates at the presynaptic nerve terminal in frog are looked for, the relatively low numbers again make it unlikely that n might represent a number of vesicles in the nerve terminal. For comparison, the numbers of vesicles in the immediate neighbourhood of the nerve membrane in the rat diaphragm synapses were counted from electron micrographs and found to be around $200/\mu^2$ (*Jones* and *Kwanbunbumpen*, 1970). Synaptic vesicles are clustered around special sites at the nerve terminal membrane. There is reason to believe that transmitter release takes place at these "active release sites" by exocytosis from synaptic vesicles (*Couteaux* and *Pecot-Dechavassine*, 1970; *Ceccarelli*, *Hurlbut* and *Mauro*, 1973; *Heuser* and *Reese*, 1973). On longitudinal sections through the terminal these sites are about 0.5 to 1.5 μ apart (*Birks*, *Huxley* and *Katz*, 1960). This would give around 100—200 such sites on a total terminal length of about 100—200 μ (*Katz*, 1969), which is in the same order of magnitude as the above estimated value for n for a whole synapse. The numbers calculated for n from *Katz and Miledi*'s experiments (1965 a, b) with a focal Ca electrode also would fit well enough for a rough comparison. Therefore n might represent the number of active release sites in a presynaptic region. A simple binomial model would require that such a site releases not more than one quantum per action potential. This does not seem impossible since the total number of quanta released per impulse at an unblocked neuromuscular junction seems to be 100 or more (*del Castillo* and *Katz*, 1954; *Martin*, 1955). This again is within the same order of magnitude as the number of active release sites per terminal and the calculated values for n. In this model p would depend on the probability that a site releases a nearby vesicle. Under usual conditions of release, entry of Ca ions into the nerve terminal in the course of depolarization (*Katz* and *Miledi*, 1967 a, b, c, 1968; see also *Baker, Hodgkin* and *Ridgway*, 1971) presumably leads to an increase in p. Before such speculations can rise above the level of a helpful working hypothesis many more experimental results on the dynamics of p and n in different experimental conditions would be needed.

Acknowledgements

Some of the work reported here was supported by the Fonds zur Förderung der wissenschaftlichen Forschung, Austria.

References

Atwood, H. L., and *G. D. Bittner:* Matching of excitatory and inhibitory inputs to crustacean muscle fibers. J. Neurophysiol. *34*, 157—170 (1971).

Atwood, H. L., and *H. S. Johnston:* Neuromuscular synapses of a crab motor axon. J. exp. Zool. *167*, 457—470 (1968).

Atwood, H. L., and *I. Parnas:* Synaptic transmission in crustacean muscles with dual motor innervation. Comp. Biochem. Physiol. *27*, 381—404 (1968).

Auerbach, A. A., and *M. V. L. Bennett:* Chemically mediated transmission at a giant fiber synapse in the central nervous system of a vertebrate. J. gen. Physiol. *53*, 183—210 (1969).

Baker, P. F., *A. L. Hodgkin*, and *E. B. Ridgway:* Depolarization and Ca entry in squid giant axons. J. Physiol. *218*, 709—756 (1971).

Birks, R., *H. E. Huxley*, and *B. Katz:* The fine structure of the neuromuscular junction of the frog. J. Physiol. *150*, 134—144 (1960).

Bittner, G. D.: Differentiation of nerve terminals in the crayfish opener muscle and its functional significance. J. gen. Physiol. *51*, 731—758 (1968).

Bittner, G. D., and *J. Harrison:* A reconsideration of the Poisson hypothesis for transmitter release at the crayfish neuromuscular junction. J. Physiol. *206*, 1—23 (1970).

Bittner, G. D., and *D. Kennedy:* Quantitative aspects of transmitter release. J. cell. Biol. *47*, 585—592 (1970).

Blackman, J. G., B. L. Ginsborg, and *C. Ray:* On the quantal release of the transmitter at a sympathetic synapse. J. Physiol. *167*, 402—415 (1963).

Blackman, J. G., and *R. D. Purves:* Intracellular recordings from ganglia of the thoracic sympathetic chain of the guinea-pig. J. Physiol. *203*, 173 to 198 (1969).

Boyd, I. A., and *A. R. Martin:* The end-plate potential in mammalian muscle. J. Physiol. *132*, 74—91 (1956).

Bracho, H., and *R. K. Orkand:* Effect of calcium on excitatory neuro-muscular transmission in the crayfish. J. Physiol. *206*, 61—71 (1970).

Ceccarelli, B., W. P. Hurlbut, and *A. Mauro:* Turnover of transmitter and synaptic vesicles at the frog neuromuscular junction. J. cell Biol. *57*, 499 to 524 (1973).

Christensen, B. N., and *A. R. Martin:* Estimates of probability of transmitter release at the mammalian neuromuscular junction. J. Physiol. *210*, 933 to 945 (1970).

Couteaux, R., and *M. Pecot-Dechavassine:* Vesicules synaptiques et poches au niveau des "zones actives" de la jonction neuromusculaire. C.R. Acad. Sci. Paris *271*, 2346—2349 (1970).

Del Castillo, J., and *B. Katz:* Quantal components of the end-plate potential. J. Physiol. *124*, 560—573 (1954).

Del Castillo, J., and *B. Katz:* Localization of active spots within the neuro-muscular junction of the frog. J. Physiol. *132*, 630—649 (1956).

Dudel, J.: Potential changes in the crayfish motor nerve terminal during repetitive stimulation. Pflügers Arch. ges. Physiol. *282*, 323—337 (1965).

Dudel, J., and *S. W. Kuffler:* Mechanism of facilitation at the crayfish neuro-muscular junction. J. Physiol. *155*, 530—542 (1961).

Elmqvist, D., and *D. M. J. Quastel:* Presynaptic action of hemicholinium at the neuromuscular junction. J. Physiol. *177*, 463—482 (1965 a).

Elmqvist, D., and *D. M. J. Quastel:* A quantitative study of end-plate potentials in isolated human muscle. J. Physiol. *178*, 505—529 (1965 b).

Fatt, P., and *B. Katz:* Spontaneous subthreshold activity at motor nerve endings. J. Physiol. *117*, 109—128 (1952).

Ginsborg, B. L.: The vesicle hypothesis for the release of acetylcholine. In: Excitatory Synaptic Mechanisms, Proceedings of the Fifth International Meeting of Neurobiologists (*Andersen, P.*, and *J. K. S. Jansen*, eds.). Oslo: Universtetsforlaget. 1970.

Heuser, J. E., and *T. S. Reese:* Evidence for recycling of synaptic vesicle membrane during transmitter release at the frog neuromuscular junction. J. cell Biol. *57*, 315—344 (1973).

Hubbard, J. I.: Mechanism of transmitter release. Progr. Biophys. Mol. Biol. *21*, 33—124 (1970).

Ishii, Y., S. Matsumura, and *T. Furukawa:* Quantal nature of transmission at the synapse between hair cells and eighth nerve fibers. Japan J. Physiol. *21,* 79—89 (1971).

Johnson, E. W., and *A. Wernig:* The binomial nature of transmitter release at the crayfish neuromuscular junction. J. Physiol. *218,* 757—767 (1971).

Jones, S. F., and *S. Kwanbunbumpen:* The effects of nerve stimulation and hemicholinium on synaptic vesicles at the mammalian neuromuscular junction. J. Physiol. *207,* 31—50 (1970).

Katz, B.: The Release of Neural Transmitter Substances. Springfield, Ill.: Thomas. 1969.

Katz, B., and *R. Miledi:* The measurement of synaptic delay and the time course of acetylcholine release at the neuromuscular junction. Proc. R. Soc. B *161,* 483—495 (1965 a).

Katz, B., and *R. Miledi:* The effect of temperature on the synaptic delay at the neuromuscular junction. J. Physiol. *181,* 656—670 (1965 b).

Katz, B., and *R. Miledi:* The release of acetylcholine from nerve endings by graded electric pulses. Proc. R. Soc. B *167,* 23—38 (1967 a).

Katz, B., and *R. Miledi:* The timing of Ca action during neuromuscular transmission. J. Physiol. *189,* 535—544 (1967 b).

Katz, B., and *R. Miledi:* Tetrodotoxin and neuromuscular facilitation. J. Physiol. *195,* 481—492 (1968).

Kuno, M.: Quantum aspects of central and ganglionic synaptic transmission in vertebrates. Physiol. Rev. *51,* 647—678 (1971).

Kuno, M., and *J. N. Weakly:* Quantal components of the inhibitory synaptic potential in spinal motoneurones of the cat. J. Physiol. *224,* 287—303 (1972).

Martin, A. R.: A further study of the statistical composition of the end-plate potential. J. Physiol. *130,* 114—122 (1955).

Martin, A. R.: Quantal nature of synaptic transmission. Physiol. Rev. *46,* 51—66 (1966).

Takeuchi, A., and *N. Takeuchi:* Active phase of frog's end-plate potential. J. Neurophysiol. *22,* 395—411 (1959).

Wernig, A.: Changes in statistical parameters during facilitation at the crayfish neuromuscular junction. J. Physiol. *226,* 751—759 (1972 a).

Wernig, A.: The effects of Calcium and Magnesium on statistical release parameters at the crayfish neuromuscular junction. J. Physiol. *226,* 761 to 768 (1972 b).

Wernig, A.: Calculations of statistical release parameters from intracellular recordings on frog neuromuscular junction. Pflügers Arch. ges. Physiol. *343,* R 72 (1973).

Zucker, R. S.: Changes in the statistics of transmitter release during facilitation. J. Physiol. *229,* 787—810 (1973).

Author's address: Prof. Dr. *A. Wernig,* Pharmakologisches Institut, Universität Innsbruck, Peter-Mayr-Straße 1, A-6020 Innsbruck, Austria.

Discussion

Ratzenhofer: The speaker has based his interesting calculations and considerations regarding the liberation of transmitter from the vesicles on exocytosis and, if I have understood correctly, on the assumption that in this process the entire contents of the vesicle are excreted into the synaptic cleft. I do not know whether the speaker's deductions can be applied to all synapses. According to *Pfenninger*'s (*see Akert*, 1971) subtile studies, in the central nervous system the vesicle contents are emptied into the cleft through the presynaptic membrane via fine synaptopores. Question: Is it known whether with this mechanism the entire contents of the vesicle or only part of them are excreted? Can your calculations also be applied to this type of transmitter liberation?

[Lit.: *Akert, K.:* Klin. Wschr. *49*, 509 (1971).]

Wernig: There is considerable evidence from many synapses, that transmitter is released in packages. To my knowledge there is as yet no electrophysiological evidence that transmitter can at all be released other than in quantal form. Unfortunately we gather all results on transmitter release from evoked electrical responses of the postsynaptic cell. This precludes any statement about other than electrically active substances which might in addition be released from the presynaptic nerve. On the other hand statistical analysis of transmitter-release seems applicable regardless of the mechanism of release, and regardless of other substances present, as long as release appears to be quantal in nature.

Andres: We have searched thoroughly in experiments with the motor end plate for the opening of the synaptic vesicles into the synaptic space. The opening of the vesicle on the motor end plate is very seldom seen because of the large surface area one must consider. I think that the model of Akert has been worked out primarily for the Gray type II synapses, since certainly other relationships are present here than at the motor end plate.

Wernig: I understand that exocytosis pictures at frog neuromuscular junction can be observed relatively seldom, which, however, seems statistically justified (*Clark, Hurlbut* and *Mauro*, 1972). If transmitter is released by exocytosis it probably makes a difference whether at the moment of fixation the nerve terminal was in "rest" or in an activated state. Recent results seem to indicate that exocytosis at the frog neuromuscular junction takes place at special sites in the nerve membrane, which for this reasons have been called "active zones" (*Couteaux et al.*, 1971; *Heuser et al.*, 1973; *Ceccarelli et al.*, 1973; see also *Couteaux* and *Pécot-Dechavassine*, 1974; *Dreyer, Peper, Akert, Sandri* and *Moor*, 1973).

[Lit.: *Clark, A. W., W. P. Hurlbut,* and *A. Mauro:* Changes in the fine structure of the neuromuscular junction of the frog caused by black widow spider venom. J. Cell Biol. *52*, 1—14 (1972); *Couteaux, M. R.,* and *M. Pécot-Dechavassine:* Les zones spécialisées des membranes présynaptiques. C.R. Acad. Sci. (Paris) *278*, 291—293 (1974); *Dreyer, F., K. Peper, K. Akert, C. Sandri,* and *H. Moor:* Ultrastructure of the "active Zone" in the frog neuromuscular junction. Brain Res. *62*, 373—380 (1973).]

Journal of Neural Transmission, Suppl. XII, 75—95 (1975)
© by Springer-Verlag 1975

Radioautography as a Tool for the Study of Putative Neurotransmitters in the Nervous System

C. Sotelo

Laboratoire de Neuromorphologie (U-106 INSERM), Hôpital de Port Royal, Paris, France

With 10 Figures

The complete demonstration that the already known neuro-transmitters in some synapses of the nervous system of invertebrates or in the peripheral nervous system of vertebrates play a similar role in the mammalian central synapses is still missing. Nevertheless the nature of the neurotransmitter present in some central pathways is well established: for instance GABA seems to be the neurotransmitter in the projections from the cerebellar cortex to the deep cerebellar nuclei and the vestibular nuclei (*Obata* and *Takeda,* 1969; *Fonnum et al.,* 1970). However, there are only a few central specific synapses in which the exact nature of the neurotransmitter is completely demonstrated (*i.e.* the cholinergic nature [*Eccles et al.,* 1954] of the synapses on Renshaw cells in the cat spinal cord).

In order to identify the substance released by a given kind of synapses arising from the same nervous pathway, certain criteria must be fulfilled. 1. The most important one is the demonstration that the specific enzyme for the synthesis of the suspected neuro-transmitter is present in the axon terminals belonging to such a pathway. There is now the possibility of a morphological approach to this problem, since immunohistochemical methods for choline acetyl transferase, glutamic acid decarboxylase and dopamine-β-hydroxylase are presently developed in several laboratories. 2. Another criterion is the proof that the suspected neurotransmitter is present in high concentration in the axon terminals under study. For some of the neurotransmitters, histochemical methods have been developed, allowing, at the light microscopic level, their direct visualization. This is the case for catecholamines and indolamines, since they can be visualized with the fluorescence technique of *Falck* and *Hillarp*. A less specific morphological approach can be used to identify neurons and axon terminals containing a given neurotransmitter. This method is the radioautography using tritiated putative neurotransmitters. It is based on the specificity of the uptake and storage mechanisms present in every system of neurons. This kind of approach will be the topic of the present communication.

It is well known that monoaminergic as well as indolaminergic neurons have the ability to take up and retain exogenous monoamines (*Glowinski et al.*, 1965). Such uptake is probably related, as postulated by *Iversen* (1967), to the inactivation mechanism, since it is well known that such inactivation occurs intracellularly. A similar uptake and neuronal concentration have also been described for other putative neurotransmitters. Thus, biochemical studies on the subcellular distribution of endogenous and ^{3}H-GABA in the rat cerebral cortex have shown that the exogenous ^{3}H-GABA can be taken up by presumptive GABA-containing neurons (*Neal* and *Iversen*, 1969) and stored into the endogenous GABA pool. Furthermore, in slices of rat cerebral cortex incubated "in vitro" in presence of ^{3}H-GABA, only this amino acid is accumulated and not its labeled metabolites (*Iversen* and *Neal*, 1968). All these results constitute propitious conditions for using radioautography to localize monoaminergic as well as GABA-containing neurons in the nervous system.

There are numerous recent reviews on the application of radioautography as a useful histochemical method for the study of putative neurotransmitters in the nervous system (*Taxi*, 1969; *Hökfelt* and *Ljungdahl*, 1972 a; *Hökfelt* and *Ljungdahl*, 1972 b; *Bloom*, 1973). Only some of the results obtained in our laboratory will be discussed here.

Localization of Catecholamines

A. Peripheral Nervous System

Since 1962, *Wolfe et al.* have shown that some water soluble substances, such as ^{3}H-noradrenaline (^{3}H-NA) can be useful tracers for radioautography when they are bound to specific storage

---

Fig. 1. Proximal segment of a sciatic nerve of the rat ligated 22 hours before the ^{3}H-NA administration. The sciatic nerve was fixed within 30 min after the labeled administration. Only some of the unmyelinated axons in the Remak fiber are labeled attesting the high specificity of the reaction. The labeling allows the identification of noradrenergic fibers. ×22,000

Fig. 2. Proximal segment of a sciatic nerve of the rat. This radioautograph illustrates the labeling obtained 3 1/2 hours after migration of ^{3}H-NA on one axonal profile. ×20,000

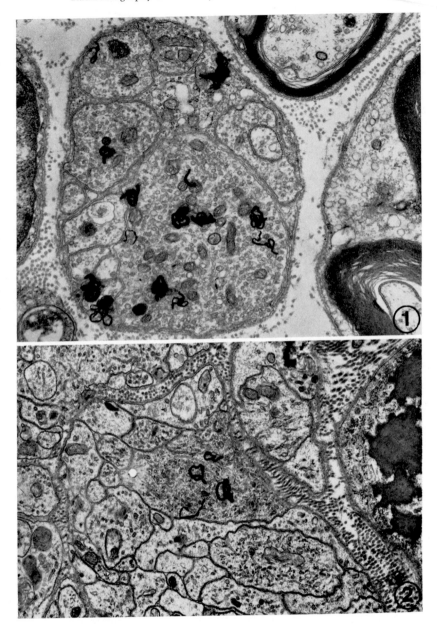

molecules, in all likehood of protein nature. The cytochemical study with high resolution radioautography of the uptake and the storage of ^{3}H-NA in the peripheral sympathetic system, the kinetics of ^{3}H-NA and its depletion by reserpine has been extensively made by *Taxi* (1969) and *Taxi* and *Droz* (1969). These authors have confirmed in normal rats the close relationship between the radioautographic labeling and the presence of small granulated vesicles (SGV) in axon terminals. In reserpinized rats, the labeling follows a parallel decrease with the disappearance of the dense core in the SGV. Thus, four hours after the administration of reserpine (5 mg/kg i.p.) there is an almost total absence of labeling and a simultaneous depletion of the dense core in the SGV (*Taxi,* 1969). These results fully validate the use of radioautography to identify monoaminergic neurons. Indeed, this technique has been most useful in our study of the axonal migration of catecholamines in the ligated sciatic nerve of the rat (*Sotelo* and *Taxi,* 1973; *Taxi* and *Sotelo,* 1973). The proximal stump of the sciatic nerve 22 hours after ligation is composed of numerous altered axonal profiles, some of them being unmyelinated. The radioautography has been the only reliable morphological method to identify those between the unmyelinated altered axonal profiles which belong to monoaminergic neurons (Fig. 1).

Radioautography of sections of the proximal stump of sciatic nerves of rats pretreated with a monoamine oxidase inhibitor (IMAO) in which the ligation has been done 20 hours before the ^{3}H-NA administration and with K Mn O4 used as primary fixative have given the following results: numerous enlarged axonal profiles contain abundant SGV, only some of them exhibit a radioautographic reaction. However, occasionally, altered axonal profiles almost only filled with agranular vesicles have a radioautographic labeling. This result may be interpreted as if the primary fixation with K Mn O4, very useful in giving a high electron density to the central core of SGV in normal monoaminergic terminals, has lost part of its specificity in the pathological conditions of the ligation.

Fig. 3. Light micrograph of a radioautogram obtained from the substantia nigra of the rat in which, after monoamine oxidase inhibition, ^{3}H-NA was intraventricularly injected. Three neuronal perikarya are heavily labeled (arrows). Unlabeled neurons are also present (N). ×1000

Fig. 4. Electron microscopic radioautograph of a dopaminergic neuron in the substantia nigra. Same experimental conditions as in Fig. 3. The nucleus as well as the cytoplasm are labeled. The peripheral cytoplasm is occupied by Nissl bodies (NB). The Golgi apparatus and its vesicular population (arrows) are practically unlabeled. ×15,000

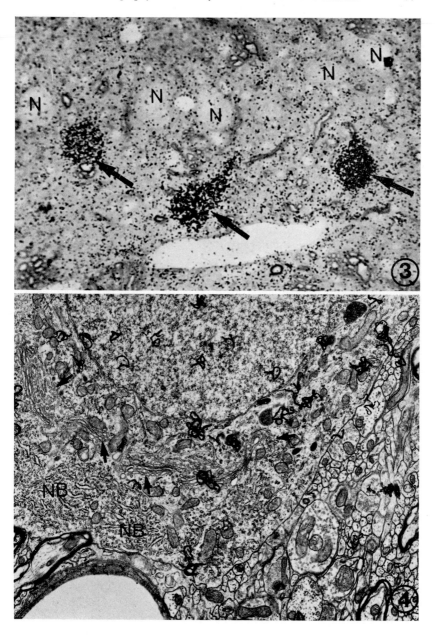

Due to the dynamic properties of the radioautographic method we have been able to analyze to what extend the axonal transport contributes to the large accumulation of noradrenaline (NA) observed with fluorescence histochemistry (*Dahlström*, 1965) in the proximal segment of sympathetic fibers in the ligated sciatic nerve of the rat. The following experiments were done (a detailed description of the methods used for this study has been published in *Sotelo* and *Taxi*, 1973): 5 mCi of ^{3}H-NA or of a precursor (^{3}H-DOPA or ^{3}H-DA) were injected intravenously into 80—100 g rats, in order to load with the tritiated monoamine the noradrenergic neurons. After a delay of 2 hours, necessary to decrease the amount of ^{3}H-NA in the blood to a negligible level, the sciatic nerves were ligated. In the rats in which the sciatic nerves were fixed 20 hours after the ligation no labeling was present in the proximal segment of sympathetic fibers. However, the proximal segment of noradrenergic sciatic nerve fibers fixed only 3 hours after the ligation exhibit a moderate radioautographic reaction (Fig. 2). The disappearance of the labeling between 3 and 20 hours in our experiments can be used as a proof that the bulk of fluorescent NA visible after 24 hours of ligation cannot be accounted for by the axonal migration of perikaryal NA, but must be related to local synthesis and storage. Similar results were obtained by *Geffen et al.* (1971) in different experiments in the splenic nerve of the cat, but also using the dynamic properties of the radioautographic method. In conclusion, the results obtained with the radioautographic method tend to prove that the axonal migration of catecholamines is only an epiphenomenon related to the distal migration of enzymatic and storage proteins from the perikaryon towards the terminals of the sympathetic axons.

B. Central Nervous System

We have used the radioautography to identify the dopaminergic neurons in the substantia nigra of the rat and monoaminergic neurons and axon terminals in the area postrema (*Sotelo*, 1971). *Dahlström* and *Fuxe* (1964), with the use of the histochemical fluorescence method for monoamines, have demonstrated the existence of medium fluorescent cells in the substantia nigra, mostly in "pars compacta". Since, it is supposed that not all the neurons present in the substantia nigra are dopaminergic, the radioautography, after treatment with ^{3}H-labeled monoamines, allows the ultrastructural recognition of the monoamine containing neurons, facilitating the cytological study of such nerve cells.

For this study, rats pretreated with an IMAO were mounted in a stereotaxic head-holder and injected in the left lateral cerebral ventricle during 3—4 min with 400 μCi of ^{3}H-NA (a detailed description of the method used for these experiments is published elsewhere, *Sotelo*, 1971). The animals were intracardially perfused with a double-aldehyde fixative 30 min after the label administration. Light microscopy radioautograms show that few scattered neurons of the "pars reticulata" have retained the ^{3}H-NA, whereas in the "pars compacta" more numerous neurons were labeled (Fig. 3).

In high resolution radioautography, silver grain clusters were located only over neuronal cell bodies (Figs. 4 and 5) and dendritic profiles (Fig. 6). None of the observed grain clusters was overlying axon terminals; even if residual radioautographic activity, represented by scattered silver grains, was associated with all the cellular elements present in the substantia nigra (glial processes, myelinated and unmyelinated axons, axon terminals, endothelial cells). In a fine structural study of the substantia nigra of the rat, done according to the neuronal size, *Gulley* and *Wood* (1971) have identified three types of nigral neurons. The medium-sized neurons, present mainly in the "pars compacta", have a diameter of 18 to 30 μm and are characterized by: 1. the presence of a rounded and almost smooth eccentric nucleus; 2. a well-developed Golgi apparatus located in a perinuclear position and projecting into the stem of primary dendrites; and 3. a well-organized rough endoplasmic reticulum, forming large masses of Nissl bodies in a peripheral position. In our radioautographic study (*Sotelo*, 1971) it is demonstrated that such neurons are those which retain the labeled catecholamine, therefore, being identified as the dopaminergic neurons (Figs. 4 and 5).

The ultrastructural study of the dopaminergic neurons has failed, till now, to reveal the presence of vesicular profiles other than those of the Golgi region. In the labeled neurons the silver grains were distributed at random. No particular cell organelle, especially the Golgi region, exhibited a specific affinity for the labeling (Figs. 4 and 5). Silver grains overlaid all cell organelles, and even the nucleus (Fig. 4) and its nucleolus were heavily labeled. This kind of distribution could be attributed to the pretreatment with an IMAO, and then it may be argued that the silver grains do no overlay the endogenous dopaminergic pool; however, *Descarries* and *Droz* (1970) have shown a similar distribution of the radioactivity in nigral neurons of non-pretreated rats.

It has been postulated that catecholamines present in monoaminergic neuronal perikarya are also stored in vesicles (*Dahlström* and *Fuxe*, 1964). The noradrenergic neurons of the sympathetic ganglia are the only well-known morphological example of a

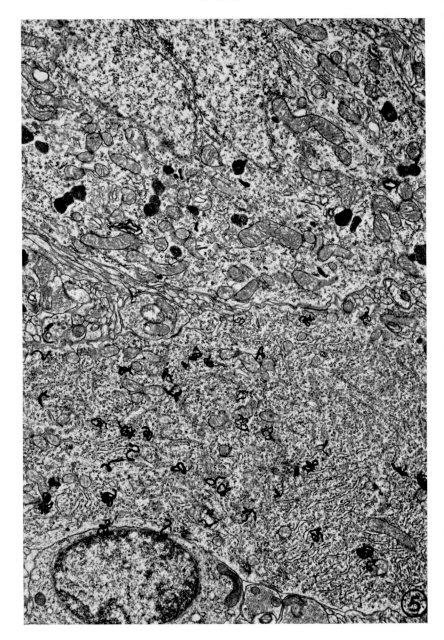

perikaryal and/or dendritic vesicular storage for catecholamines (*Taxi*, 1965; *Grillo*, 1966). Without going to the discussion of the functional role played by such vesicles (see, for instance, in *Taxi et al.*, 1969), it is certain that these clusters of vesicles can be specifically labeled even in rats not pretreated with an IMAO (*Taxi*, in preparation). Therefore, radioautographic results discussed here for the nigral dopaminergic neurons, as well as those reported by *Descarries* and *Droz* (1970), do not support the hypothesis that monoamines in these neurons are also stored in vesicles. The present results are in agreement with the hypothesis advanced by *Descarries* and *Droz* (1970) that in catecholaminergic neurons a macromolecular complex of probable protein nature would be synthetized in the perikaryon and could be the binding site for the monoamines. This nonvesicular binding material must be present throughout the monoaminergic neuron, its perikaryon and processes (Figs. 4, 5 and 6).

In the area postrema, besides the presence of only rare neurons exhibiting a radioautographic reaction of weaker intensity than the nigral neurons, axon terminals were also labeled. These labeled terminals were scattered in the neuropil; none of them established synaptic contacts with other neuronal perikarya, but only with small dendritic profiles. The labeled boutons were mainly filled with agranular vesicles (Fig. 7), and only occasionally were large granular vesicles (LGV) present in them. Therefore, the presence of LGV in central axon terminals is not proof that such terminals should be identified as monoaminergic, although in some instances, LGV have been demonstrated to be storage organelles for central monoamines (*Richards* and *Tranzer*, 1970).

GABA-Containing Neurons

Numerous biochemical, electrophysiological and neuropharmacological studies have been published in favour of the important role of GABA as an inhibitory neurotransmitter substance in the CNS. In order to achieve a better knowledge on the cellular localization of this amino-acid in the CNS radioautography can be very useful. Indeed, after the papers by *Bloom* and *Iversen* (1971) and *Iversen* and *Bloom*

Fig. 5. Electron microscopic radioautograph of the substantia nigra. Same experimental conditions as in Fig. 3. Two nigral neurons are illustrated in the micrograph. The one at the top is unlabeled and corresponds to the small type of neuron described by *Gulley* and *Wood* (1971). The one at the bottom is labeled. It is characterized by its medium-size and mainly by the large masses of Nissl bodies present in the cytoplasm. This figure clearly illustrates that the dopaminergic neurons correspond to the medium-sized type of *Gulley* and *Wood* (1971). Picture from *Sotelo* (1971). ×15,000

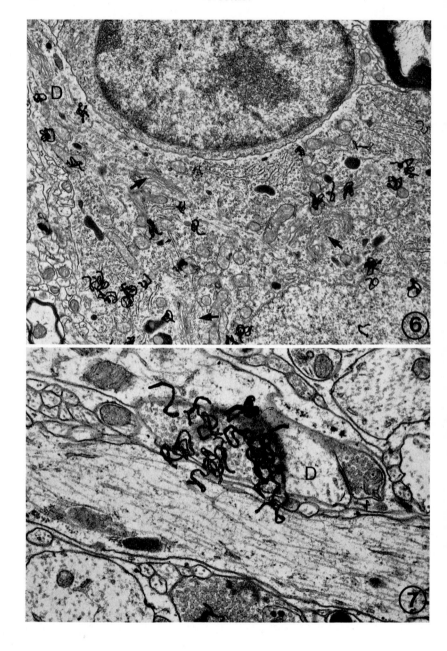

(1972) it was shown that ^{3}H-GABA is preferentially taken up by nerve terminals and that the radioautographic localization of the ^{3}H-GABA may be useful to identify those neurons using GABA as neurotransmitter.

Since ^{3}H-GABA does not cross the blood-brain barrier, here again the way of administration is crucial to obtain a specific radioautographic reaction. We have used, in non-treated rats, injection into the cerebrospinal fluid "in vivo" of 100 μCi of ^{3}H-GABA in 20 μl saline. The rats were sacrificed 30 min after the injection. In none of the regions in which we looked for radioautographic reaction (cerebral cortex, cerebellum, substantia nigra) has such reaction been observed. More recently, *Schon* and *Iversen* (1972) have used the same intraventricular administration, but in rats pretreated with aminooxyacetic acid (AOAA) to prevent the fast metabolism of ^{3}H-GABA. Under these conditions, these authors were able to label suspect GABA-containing neurons in the cerebellar cortex; but, due to the limited diffusion of ^{3}H-GABA from the ventricular system to the nervous parenchyma, only the most superficial stellate cells were labeled.

The incubation of slices of different regions of mammalian CNS, "in vitro" with ^{3}H-GABA have been used for radioautographic purposes (*Hökfelt* and *Ljungdahl*, 1970; *Bloom* and *Iversen*, 1971; *Iversen* and *Bloom*, 1972; *Hattori et al.*, 1973). This "in vitro" technique generally gives a very poor morphology. Furthermore, the specific uptake mechanism of GABA-containing neurons may be altered by damaging neuronal perikarya or processes during the dissection and the cutting procedures.

A third method of ^{3}H-GABA administration is intracerebral injection. We have mainly used this third approach in our study of suspected GABA-containing axon terminals in the substantia nigra of the rat. The method used has been already described (*Agid et al.*, 1973) in a previous paper.

Light microscopic study of radioautograms obtained from the substantia nigra injected directly with ^{3}H-GABA shows that silver grain clusters are numerous in the neuropil of the region around the tip of the injection cannula. The radioautographic picture is just

Fig. 6. Electron microscopic radioautograph of the substantia nigra. Same experimental conditions as in Fig. 3. A medium-sized neuron exhibits a similar density of labeling in the perikaryon and in the stem of a dendrite (D). The arrows point to the unlabeled Golgi vesicular population. $\times 15,000$

Fig. 7. Electron microscopic radioautograph of the area postrema. Same experimental conditions as in Fig. 3. One axon terminal in synaptic contact with a small dendritic profile (D) and containing numerous agranular flattened vesicles is heavily labeled. Picture from *Sotelo* (1971). $\times 30,000$

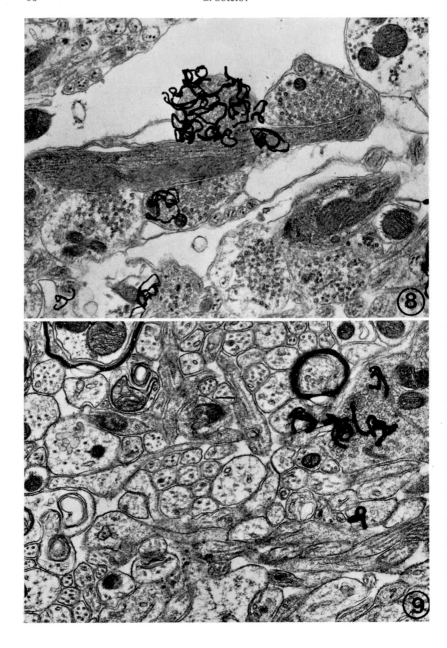

opposite to that obtained after intraventricular administration of
^{3}H-NA. In the latter case, as described above, the silver grain clusters
were only overlaying neuronal cell bodies and dendrites. In the
^{3}H-GABA intraparenchymal administration, the grain clusters were
spread in the neuropil and over glial perikarya. Most of the significant
radioautographic reaction was circumscribed to the edematous zone
of the injection and to a narrow band around this edematous area.
Glial labeling is observed far from the injected area. Thus, injections
placed in the "pars compacta" provide a significant radioautographic
reaction in an area 200 to 300 μm in diameter; however, intense
labeling is present in the astrocytic processes of the "crus cerebri",
till the subpial glial lamina. These results prove that, although
^{3}H-GABA can spread from long distances in the brain parenchyma,
only small zones, near the injection site, can be used to study the
GABA-containing neurons. These zones been mostly the edematous
areas produced around each injection site. The present observations
may invalidate the use of this kind of approach to establish the
cartography of GABA-containing neurons in a given neuronal nucleus
or region, since the site of labeling is tributary of the localization of
the tip of the injection cannula.

One important reason to explain the present results is the inter-
vention of glial cells in the uptake of ^{3}H-GABA in the presence of
an excess of this amino acid in the extracellular space. This glial
uptake may be important to protect the neuronal membranes from
being in contact with the excess of extracellular GABA. Radioautographic studies have proven the glial uptake. Thus, *Neal* and *Iversen*
(1972) have shown that in the isolated retina ^{3}H-GABA uptake
mostly occurs into the Müller cells. These kind of studies have been
recently very useful to understand some of the discrepant results
obtained by *Bowery* and *Brown* (1972). These investigators have
repeated the experiments of *Iversen* and *Neal* (1968), using, instead
of slices of cerebral cortex, isolated sympathetic ganglia, a material
which seems devoid of GABA neurons. They have reported an uptake

Fig. 8. Electron microscopic radioautograph of the substantia nigra of a rat
pretreated with AOAA and stereotaxically injected in "pars compacta" with
^{3}H-GABA. An axo-dendritic synaptic bouton is heavily labeled. This axon terminal
is located in a zone with a severe extracellular edema, indicating its proximity to
the site of the injection. Isolated silver grains are indicative of the diffuse reaction
probably due to the presence of free ^{3}H-GABA. $\times$ 25,000

Fig. 9. Electron microscopic radioautograph of substantia nigra. Same experimental
condition as in Fig. 8. A GABA-containing axon terminal is labeled at the right of
the micrograph. The extracellular edema is practically absent, indicating that this
labeled bouton is not near the injection site. $\times$ 40,000

and accumulation phenomena similar to that occuring in the cerebral cortex. Furthermore, this ^{3}H-GABA can be released from the ganglia. Recently, *Young et al.* (1973) have repeated similar experiments and studied the localization of ^{3}H-GABA with the radioautographic method. They have demonstrated that most of the ^{3}H-GABA accumulated by the sympathetic ganglia is taken up by the glial cells, mainly by the satellite gliocytes. The glial uptake was corroborated and the results of *Bowery* and *Brown* (1972) explained.

Concerning the GABA innervation of the rat substantia nigra, the ultrastructural examination of radioautograms reveals that the dense aggregates of silver grains mainly correspond to axon terminals. In zones with intense extracellular edema, indicating their proximity to the injection points, some axon terminals in axo-dendritic synaptic contact present a high radioautographic reaction (Fig. 8). In these zones, the neuropil surrounding the labeled terminals exhibits a ubiquitous diffuse reaction. This diffuse reaction, illustrated in Fig. 8, corresponds to the sparse silver grains observed almost over all cellular elements present in the neuropil. In the narrow band around the edematous area, in which the nervous processes are better preserved, few axon terminals dispersed in the neuropil also exhibit an intense accumulation of silver grains (Fig. 9).

The use of intracerebral injections of ^{3}H-GABA allows the radio-autographic examination of only small areas of the brain located around the tip of the injection cannula. Furthermore, in the "pars compacta" of the substantia nigra only few axon terminals exhibit a real accumulation of silver grains (4 to 10 silver grains per axon terminal profile). A sparse reaction, interpreted here as being due to the presence of free ^{3}H-GABA and identified by the existence of isolated silver grains, is labeling a large portion of the axonal population. These results are quantitatively different from those of *Hattori et al.* (1973) using incubation of slices of substantia nigra, or those reported by *Iversen et al.* (1973) also with the same approach. For the latter investigators more than 50 % of nerve terminals are capable of accumulating ^{3}H-GABA. This discrepancy may be due to the technical differences (incubation of slices versus intracerebral injections) or to the fact, as in the *Hattori et al.* electron micrographs, that one single silver grain is taken into consideration as a specific marker of GABA-containing terminals.

A last point of this review will be concerned with the GABA-neurons present in the cerebellar cortex, since a large amount of circumstantial evidence has been accumulated during the last years indicating that GABA probably is the neurotransmitter of the Purkinje cells as well as of the other inhibitory interneurons of the

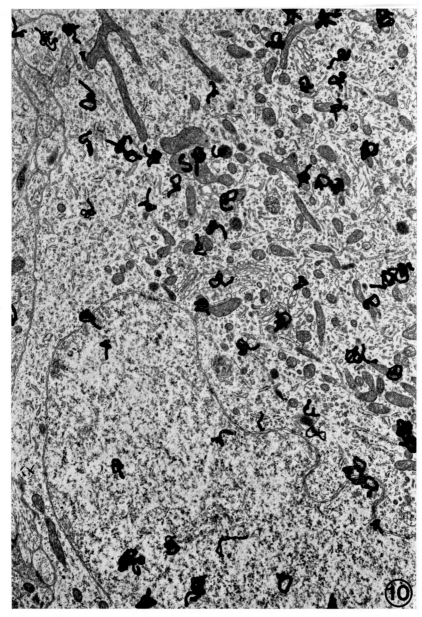

Fig. 10. Electron microscopic radioautograph of an immature Purkinje cell taken from a tissue culture of rat cerebellum maintained for 7 days "in vitro" and incubated for 30 min with ^{3}H-GABA. The silver grains are randomly distributed over the nucleus and the cytoplasm. The Purkinje cell is devoid of glial envelope.

×15,000

mammalian cerebellum (*Roberts* and *Kuriyama*, 1968; *Obata* and *Takeda*, 1969; *Fonnum et al.*, 1970; *Kawamura* and *Provini*, 1970; *Bisti et al.*, 1971). The three different ways of ³H-GABA administration have been used to study the cerebellar GABA neurons. The intraventricular injection, even in animals pretreated with AOAA, only allows, as mentioned before, the labeling of the most superficial stellate cells (*Schon* and *Iversen*, 1972). The "in vitro" incubation of cerebellar slices (*Hökfelt* and *Ljungdahl*, 1970) and the intracerebral injection (*Hökfelt* and *Ljungdahl*, 1972 c) allow the labeling of the cerebellar inhibitory interneurons. In none of these instances were the Purkinje cells, the most sure candidate as GABA-containing neuron (*Obata* and *Takeda*, 1969; *Fonnum et al.*, 1970), labeled. However, using another approach, the tissue culture of rat cerebellum in which the labeled amino-acid can be directly added to the culture medium, we have been able to label immature Purkinje cells (*Sotelo et al.*, 1972). In this study, dense accumulation of silver grains were observed over small (about 7 μm in diameter), medium-sized (about 12 μm in diameter), and larger (about 15 μm in diameter) neuronal cell bodies. The electron microscopic examination of radioautograms obtained from such cultures have allow the identification of the small neurons as stellate cells, the medium-sized neurons as baskets or Golgi cells and the larger neurons as Purkinje cells. Fig. 10 illustrates one immature Purkinje cell with its typical eccentric indented nucleus, a cytoplasm with numerous free polyribosomes and only few cisterns of rough endoplasmic reticulum and crowded with mitochondria. The silver grains are distributed at random all over the cytoplasm and the nucleus. The different results concerning the Purkinje cells obtained in the "in vivo" adult cerebellum and in tissue culture may partially be due to the fact that in our tissue culture experiments (fragments of newborn rat cerebellum were explanted and cultivated for 7 days) the Purkinje cells were devoid of glial envelope. Thus, the glial GABA-uptake been absent, the ³H-GABA can reach the Purkinje cells demonstrating that these cells are able to take up and accumulate the labeled amino-acid. Recently, *Ljungdahl et al.* (1973), using another model, intraocular cerebellar transplants, have corroborated the uptake of ³H-GABA by Purkinje cells. They also suggested that the glial investment of the Purkinje cells in their cerebellar transplants was missing or incomplete.

In conclusion, the radioautographic method is a useful tool for the identification of monoamine- and/or GABA-containing neurons. Unfortunately, the specificity of this method is not as absolute as the immunohistochemical demonstration of neurotransmitter synthetizing enzymes, such as DOPA-decarboxylase, dopamine-β-hydroxylase and/or glutamic acid decarboxylase. Radioautography is based on the specificity of uptake, but in some instances, a positive uptake of a putative neurotransmitter does not necessarily mean that such molecules really act as neurotransmitters. A well known example of this process is the physiological uptake of serotonin by sympathetic terminals in the pineal gland, in which NA is the neurotransmitter. However, there are many other examples in which there exists a close correlation between the presence of the neurotransmitter synthetizing enzyme and the specific uptake of such neurotransmitters (*i.e.* for dopamine: the nigro-noestriatal pathway; for GABA: the projections from the cerebellar cortex to the vestibular nuclei etc.). Due to the existence of this correlation the radioautographic method can be used as a complementary approach to the study of neurotransmission.

References

Agid, Y., F. Javoy, J. Glowinski, D. Bouvet, and *S. Sotelo:* Injection of 6-hydroxydopamine into the substantia nigra of the rat. II. Diffusion and specificity. Brain Res. *58,* 291—301 (1973).

Bisti, S., G. Iosif, G. F. Marchesi, and *P. Strata:* Pharmacological properties of inhibitions in the cerebellar cortex. Exp. Brain Res. *14,* 24—37 (1971).

Bloom, F. E.: Ultrastructural identification of catecholamine containing central synaptic terminals. J. Histochem. Cytochem. *21,* 333—348 (1973).

Bloom, F. E., and *L. L. Iversen:* Localization of ³H-GABA in nerve terminals of rat cerebral cortex by electron microscopic autoradiography. Nature *229,* 628—630 (1971).

Bowery, N. G., and *D. A. Brown:* γ-aminobutyric acid uptake by sympathetic ganglia. Nature New Biol. *238,* 89—91 (1972).

Dahlström, A.: Observations on the accumulations of noradrenaline in the proximal and distal parts of peripheral adrenergic nerves after compression. J. Anat. (London) *99,* 677—689 (1965).

Dahlström, A., and *K. Fuxe:* Evidence for the existence of monoamine-containing neurons in the central nervous system. Acta physiol. Scand. suppl. *232,* 62, 6—55 (1964).

Descarries, L., and *B. Droz:* Intraneural distribution of exogenous norepinephrine in the central nervous system of the rat. J. Cell Biol. *44,* 385—399 (1970).

Eccles, J. C., P. Fatt, and *K. Koketsu:* Cholinergic and inhibitory synapses in a pathway from motor-axon collaterals to motoneurones. J. Physiol. (London) *216,* 524—562 (1954).

Fonnum, F., J. Storm-Mathisen, and *F. Walberg:* Glutamate decarboxylase in inhibitory neurons. A study of the enzyme in Purkinje cell axons and boutons in the cat. Brain Res. *20,* 259—275 (1970).

Geffen, L. B., L. Descarries, and *B. Droz:* Intraaxonal migration of ³H-norepinephrine injected into the coeliac ganglion of cats: radioautographic study of the proximal segment of constricted splenic nerves. Brain Res. *35,* 315—318 (1971).

Glowinksi, J., I. J. Kopin, and *J. Axelrod:* Metabolism of ³H-norepinephrine in the rat brain. J. Neurochem. *12,* 25—30 (1965).

Grillo, M. A.: C. Electron microscopy of sympathetic tissues. Pharmacol. Rev. *18,* 387—399 (1966).

Gulley, R. L., and *R. L. Wood:* The fine structure of the neurons in the rat substantia nigra. Tissue and Cell *3,* 675—690 (1971).

Hattori, T., P. L. McGeer, H. C. Fibiger, and *E. G. McGeer:* On the source of GABA-containing terminals in the substantia nigra. Electron microscopic autoradiographic and biochemical studies. Brain Res. *54,* 103 to 114 (1973).

Hökfelt, T., and *A. Ljungdahl:* Cellular localization of labeled gamma-aminobutyric (³H-GABA) in rat cerebellar cortex; and autoradiographic study. Brain Res. *22,* 391—396 (1970).

Hökfelt, T., and *A. S. Ljungdahl:* Histochemical determination of neurotransmitter distribution. In: Neurotransmitters (Res. Publ. A.R.N.M.D.) Vol. *50,* pp. 1—55 (1972 a).

Hökfelt, T., and *A. Ljungdahl:* Application of cytochemical techniques to the study of suspected transmitter substances in the nervous system. In: Studies of Neurotransmitters at the Synaptic Level (*Costa, E., Iversen, L. L.,* and *R. Paoletti,* eds.), pp. 1—35 (Advances in Biochemical Psychopharmacology, Vol.6). New York: Raven Press. 1972 b.

Hökfelt, T., and *A. Ljungdahl:* Autoradiographic identification of cerebral and cerebellar cortical neurons accumulating labeled Gamma-amino-butyric acid (³H-GABA). Exp. Brain Res. *14,* 354—362 (1972 c).

Iversen, L. L.: The uptake and storage of noradrenaline in sympathetic nerves, p. 253. Cambridge University Press. 1967.

Iversen, L. L., and *F. E. Bloom:* Studies of the uptake of ³H-GABA and ³H-glycine in slices and homogenates of rat brain and spinal cord by electron microscopic autoradiography. Brain Res. *41,* 131—143 (1972).

Iversen, L. L., J. S. Kelly, M. Minchin, F. Schon, and *S. R. Snodgrass:* Role of amino acids and peptides in synaptic transmission. Brain Res. *62,* 567—576 (1973).

Iversen, L. L., and *M. J. Neal:* The uptake of ³H-GABA by slices of rat cerebral cortex. J. Neurochem. *15,* 1141—1149 (1968).

Kawamura, H., and *L. Provini:* Depression of cerebellar Purkinje cells by microiontophoretic application of GABA and related amino acids. Brain Res. *24,* 293—304 (1970).

Ljungdahl, A., A. Seiger, T. Hökfelt, and *L. Olson:* ³H-GABA uptake in growing cerebellar tissue: autoradiography of intraocular transplants. Brain Res. *61,* 379—384 (1973).

Neal, M. J., and *L. L. Iversen:* Subcellular distribution of endogenous and ³H-γ-aminobutyric acid in rat cerebral cortex. J. Neurochem. *16,* 1245 to 1252 (1969).

Neal, M. J., and *L. L. Iversen:* Autoradiographic localization of ³H-GABA in rat retina. Nature New Biology *235,* 217—218 (1972).

Obata, K., and *K. Takeda:* Release of γ-aminobutyric acid into the fourth ventricle induced by stimulation of the cat's cerebellum. J. Neurochem. *16,* 1043—1047 (1969).

Richards, J. G., and *J. P. Trancer:* The ultrastructural localization of amine storage sites in the central nervous system with the aid of a specific marker, 5-hydroxydopamine. Brain Res. *17,* 463—469 (1970).

Roberts, E., and *K. Kuriyama:* Biochemical-physiological correlations in studies of the γ-aminobutyric acid system. Brain Res. *8,* 1—35 (1968).

Schon, F., and *L. L. Iversen:* Selective accumulation of ³H-GABA by stellate cells in rat cerebellar cortex in vivo. Brain Res. *42,* 503—507 (1972).

Sotelo, C.: The fine structural localization of norepinephrine-³H in the substantia nigra and area postrema of the rat. An autoradiographic study. J. Ultrastruct. Res. *36,* 824—841 (1971).

Sotelo, C., A. Privat, and *M. J. Drian:* Localization of ³H-GABA in tissue culture of rat cerebellum using electron microscopy radioautography. Brain Res. *45,* 302—308 (1972).

Sotelo, C., and *J. Taxi:* On the axonal migration of catecholamines in constricted sciatic nerve of the rat. A radioautographic study. Z. Zellforsch. *138,* 345—370 (1973).

Taxi, J.: Contribution à l'étude des connexions des neurones moteurs du système nerveux autonome. In: Annales des Sciences Naturelles, 12e série, tome VII, Fasc. 3, pp. 413—674. Paris: Masson Ed. 1965.

Taxi, J.: Morphological and cytochemical studies on the synapses in the autonomic nervous system. In: Progress in Brain Research, Vol. 31, pp. 17—20. Amsterdam: Elsevier. 1969.

Taxi, J., and *B. Droz:* Radioautographic study of the accumulation of some biogenic amines in the autonomic nervous system. In: Symposium of the International Society for Cell Biology, Vol. 8, pp. 175—190. New York: Academic Press. 1969.

Taxi, J., J. Gautron, and *P. L'Hermite:* Données ultrastructurales sur une éventuelle modulation adrénergique de l'activité du ganglion cervical supérieur du Rat. C. R. Acad. Sc. Paris. t. *269,* série D, 1281—1284 (1969).

Taxi, J., and *C. Sotelo:* Cytological aspects of the axonal migration of catecholamines and of their storage material. Brain Res. *62,* 431—437 (1973).

Wolfe, D. E., L. T. Potter, K. C. Richardson, and *J. Axelrod:* Localizing tritiated norepinephrine in sympathetic axons by electron microscopy autoradiography. Science *138,* 440—442 (1962).

Young, J. A. C., D. A. Brown, J. S. Kelly, and *F. Schon:* Autoradiographic localization of sites of ³H-γ-aminobutyric acid accumulation in peripheral autonomic ganglia. Brain Res. *63,* 479—486 (1973).

Author's address: Dr. *C. Sotelo,* Laboratoire de Neuromorphologie, U-106 I.N.S.E.R.M., Hôpital de Port Royal, 123, bd. de Port Royal, F-75014 Paris, France.

Discussion

Dahlström: 1. About the non-vesicular pool in central MA neurons which may of course exist: I understood that your experiment included MAO inhibition together with intraventricular administration of the amine. Under these circumstances it is not astonishing that you can see silver grains over the nucleus. Fluorescence-microscopically such treatment (with MAO inhibition) induces a raised amine fluorescence in the nucleus and all parts of the neuron contain amine fluorescence probably due to increased levels of cytoplasmic amines. The cytoplasmic amines probably diffuse also into the nucleus.

2. Under such conditions I think that any attempt to relate amine uptake to any particular organelle are doomed to fail and, since amine uptake is, as you say, impossible to study without MAO inhibition, I feel that the question of a non-vesicular (or non-granular) pool of amine in CNS catecholamine nerve cell bodies is impossible to study.

3. Also, vesicular profiles do occur, although rarely in central CA neurons, according to personal communication from Dr. Hökfelt.

Sotelo: I probably was not clear to you during my presentation. I have not said that it is impossible to study amine uptake without MAO inhibition. The results by *Descarries* and *Droz* (1970) for central monoaminergic neurons and recently by *Taxi* (illustrated during my presentation) for nor-adrenergic sympathetic ganglion cells are demonstrative enough of the fact that such uptake can be studied radioautographically in non-pretreated rats. What I have actually said is that the inhibition of the MAO highly increases the radioautographic activity, without any apparent change of the cellular localization of the silver grains (*Descarries* and *Droz,* 1970). Furthermore, in the sympathetic neurons, where a perikaryal and dendritic vesicular pool is present, the silver grains overlaid specifically the vesicular clusters (*Taxi,* unpublished observations). Such vesicular pool has not been observed in the central monoamine-containing neurons that I have studied.

Whittaker: Have you tried to localize [³H] choline by autoradiography? In my group we had no success and the method only seems to work with primary amines. Presumably these are converted to instable products by the fixative.

Sotelo: No, we have not tried. I agree that the radioautographic failure with ³H-choline is probably due to the lability, during the fixation process, of the binding of the soluble acetylcholine to an insoluble matrix.

Thoenen: I would like to ask you about the localization of silver grains in autoradiographies. In spite of the use of MAO inhibitors, it is possible that you get non-enzymatic oxidation products of dopamine which could be bound covalently to nucleophilic groups of macromolecules such as proteins and nucleic acids. Have you any information how much of the radioactivity present in your tissue samples is covalently bound *i.e.* cannot be extracted by TCA?

Sotelo: Your question is very pertinent. First at all, let me remind you that the labeling of the nucleus also occurs in non-pretreated animals

(*Descarries* and *Droz*, 1970; Fig. 9). Even without checking how much of the radioactivity is bound to the TCA precipitable substances in our material, we have explained the radioautographic reaction of the neuronal nuclei as due to an artifact (*Sotelo*, 1971). This artifact is not only present in the study of catecholamine containing neurons, but also it is present in the study of other putative transmitters such as ^{3}H-GABA. In any case, this method is reliable enough for the identification of dopamine-containing neurons in the substantia nigra.

Pilgrim: I believe it justified to assume that the substances which are retained in a section prepared for electron microscopy are approximately identical with the TCA-precitable fraction (with the exception of part of the lipids).

Journal of Neural Transmission, Suppl. XII, 97—114 (1975)
© by Springer-Verlag 1975

Intra-Axonal Transport of Transmitters in Mammalian Neurons

A. Dahlström and P. O. Heiwall

Institute of Neurobiology, University of Göteborg, Sweden

With 5 Figures

Introduction

In adrenergic as well as in cholinergic peripheral neurons a proximo-distal transport of the respective neurotransmitters, noradrenaline (NA) and acetylcholine (ACh), has been demonstrated (cf. *Dahlström* and *Häggendal*, 1966; *Dahlström et al.*, 1974 a). This transport of transmitter molecules from more proximal levels of the neuron to the nerve endings is, in all likelyhood, not necessary to maintain the transmitter levels of the nerve terminals. Since many years it is known that adrenergic (cf. *Iversen*, 1967) as well as cholinergic (cf. *Katz*, 1966) nerve terminals possess the capacity for an efficient *de novo* synthesis of NA and ACh, respectively. The NA and ACh, which is transported distally along axons, may rather be regarded as markers for transported organelles or macromolecules. These organelles or macromolecules in turn may be essential for maintaining the nerve terminal functions, as is probably the case for adrenergic neurons (cf. *Dahlström* and *Häggendal*, 1973). Some evidence have recently been obtained in support of the view that similar mechanisms may operate also in cholinergic nerves, but more data need to be obtained. The aim of this presentation is to give a short review of results on neurotransmitter transport in adrenergic and cholinergic neurons.

I. Adrenergic Neurons

The intra-axonal transport of NA in adrenergic neurons has been studied both histochemically, with the use of the Hillarp-Falck method (for ref. see *Corrodi* and *Johnsson*, 1967; *Falck* and *Owman*, 1965), and biochemically, using the trihydroxy indole method as modified by *Häggendal* (1963).

All parts of the adrenergic neurons can be studied in the fluorescence microscope, due to their content of NA. Under normal conditions, the cell bodies show a medium fluorescence intensity, while the non-terminal axons have a very weak fluorescence due to their low NA concentration. The very wide-spread nerve terminal arborizations (see *Dahlström*, 1967 a) consist of thin fibres with bag-like enlargements (varicosities) at regular intervals (Fig. 1). These varicosities are strongly fluorescent due to their very high amounts of NA, and represent the presynaptic structures from which NA is released at nerve activity. It is estimated that the added lengths of the nerve terminal branches of one peripheral adrenergic neuron can amount to between 10—30 cm and contain about 25,000 varicosities (cf. *Dahlström*, 1967 a). This is thus a very widely spread neuron system, and since only the perikarya contain the protein synthesizing and -package machineries one might expect a very active synthesis and transport distally of material needed to support and maintain the nerve terminals.

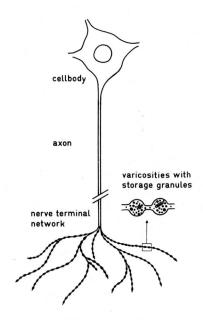

Fig. 1. Schematic drawing of a peripheral adrenergic neuron (from *Dahlström et al.*, 1974 a)

a) NA Transport

One method to investigate intra-axonal transport of a substance is to perform an axotomy, e.g. a crush, of a peripheral nerve and study the accumulation of the substance in the axons near the crush region. The sciatic nerve of e.g. rat and cat contains, in addition to the motor and sensory nerves, also a large number of thin unmyelinated adrenergic axons. With the histochemical technique it was observed that already 5—10 min after crushing a sciatic nerve of rat, NA fluorescence accumulated just proximal to the crush. The amount of NA fluorescence increased rapidly with time after crushing, indicating an interrupted proximo-distal rapid transport of NA in these neurons (*Dahlström*, 1965). A small accumulation also occurred distal to the crush, but was only short-lasting and of much less magnitude. This small retrograde accumulation may indicate a retrograde transport, but this has not yet been demonstrated to occur for NA in adrenergic neurons (cf. *Dahlström*, 1971 a). The NA accumulations could be depleted by e.g. reserpine and tetrabenzine (2 agents which block the NA storage capacity of amine storage granules) and it was therefore suggested that the NA which was transported distally in these adrenergic axons, was bound to amine storage granules (*Dahlström*, 1965, 1967 b; *Dahlström* and *Häggendal*, 1966). These granules were first isolated by *von Euler* and *Hillarp* (1956) and have been shown to store the major part of the NA in adrenergic nerve terminals, and to be essential for the release of NA at nerve activity (cf. *Andén*, *Carlsson* and *Häggendal*, 1969).

The subsequent quantitative estimations of the amount of NA which accumulated proximal to a crush indicated that the NA amount in the 1 cm part of nerve proximal to the crush rose approximately linearly with time (*Dahlström* and *Häggendal*, 1966). Assuming a) that all of the axonal NA was transportable, and b) that the amine granules did not markedly change their NA content while being transported, a transport rate of 5 mm/hour was calculated for rat sciatic adrenergic neurons. Ad a): Recent experiments indicate, however, that only about 50 % of the axonal NA is transportable (*Häggendal*, *Dahlström* and *Larsson*, 1975). The nerve segments distal to the crush were assayed for NA content 1—6 hours after operation. The NA level decreased to about 50 % and remained at this level. Since it is known that rapid proximo-distal transport continues distal to a crush or cut (cf. *Ochs* and *Ranish*, 1969), the 50 % decrease was probably due to a shift of this fraction of NA into more distal parts of the nerve. The remaining NA could probably not be transported further distally with the fast transport, and may therefore be called stationary or non-mobile NA (most of this NA is probably localized to vasoconstrictor nerve terminals in the blood vessels supplying the sciatic nerve). Ad b): It seems likely that the amine granules *do* increase their NA content while being transported and when arrested above a crush. In experiments where a second high crush was performed 6 hours after a low crush and the segment between the two crushes was assayed 1—12 hours after the *second* crush, the NA content rose significantly (*Dahlström, Häggendal* and *Larsson*, 1975). Since the number of granules could not have increased (no supply from proximal parts of the axon because of the high crush) and since all NA was reserpine depletable (*i.e.* stored in amine granules), this indicates an increased loading of NA per average amine granule of almost 100 % (*Dahlström et al.*, 1975). Taking these two pieces of information into account, when calculating the rate of transport of NA granules in rat sciatic nerve, gives a figure of about 8 mm/hour (cf. *Häggendal et al.*, 1974).

b) Transport of Enzymes

The amine granules in peripheral adrenergic neurons contain, in addition to NA, the enzyme which converts dopamine (DA) into NA, the DA-β-hydroxylase (DBH) (cf. *De Potter*, 1971). This enzyme is accumulating above a crush in parallel with the NA accumulations, which would be expected if the two substances were contained in the same subcellular unit (*Laduron* and *Belpaire*, 1968; *Livett et al.*, 1968; *Brimijoin*, 1972). The other NA synthesizing enzymes, tyrosine hydroxylase (TH) and dopa-decarboxylase (DDC) appear to be transported at rates slower than that of amine granules (*Oesch, Otten* and *Thoenen*, 1973; *Dairman et al.*, 1973), but contradictory results have been reported (*e.g. Wooten* and *Coyle*, 1973). It is possible that the results may vary because of variability of the transportable fraction in the different experiments. It is possible that *e.g.* TH may exist in both a soluble, very slowly transportable form, and a more rapidly transportable form, perhaps adsorbed to organelle membranes. If the ratio between the two forms varies in different systems, various amounts of TH accumulations above a crush can be found. Since the rate of transport in most studies is calculated on the bases of accumulated amounts compared with the total content in uncrushed nerves, various figures may be obtained despite a possible general similarity in rates for slow or rapid transport in different neurons.

c) Electron Microscopic Observations

The amine storage granules or vesicles in adrenergic neurons can be visualized in electron microscopical (E.M.) preparations. If

KMnO$_4$ fixation is used almost all vesicles, both the large type (800 to 1000 Å) and the small type (400—500 Å), have an electron dense core (*e.g. Hökfelt,* 1969). In glutaraldehyde-OsO$_4$ fixed tissues mainly the large type is dense-cored. It should be noted that with this last mentioned fixation technique the dense core does not necessarily represent the presence of a monoamine, but may rather indicate the presence of a protein matrix in the organelle. In crushed adrenergic axons a large number of mainly large dense core vesicles accumulate proximal to the crush (*Kapeller* and *Mayor,* 1966; *Geffen* and *Ostberg,* 1969). Such vesicles have also been observed in uncrushed axons and in perikarya of peripheral adrenergic neurons, but they appear to be rather sparsely distributed in the thin E.M. sections (*e.g. Hökfelt,* 1969). In the nerve terminals the majority of the vesicles are of the small size, while a varying, but rather small percentage are large (*Geffen* and *Ostberg,* 1969; *Hökfelt,* 1969). In fractionation studies of tissues containing adrenergic nerve terminals two peaks of NA have been found (*e.g. Roth et al.,* 1968; *Bisby* and *Fillenz,* 1971; *De Potter,* 1971). The more dense peak contained large dense core vesicles while in the lighter fraction small dense core vesicles were seen (*Bisby* and *Fillenz,* 1971). Since the large type seems to be the type which is transported distally in the axons, it is possible that it may represent a "young" type of vesicle, which gives rise to the small vesicle. Evidence for a functional heterogeneity of amine storage particles exist, and indications that the "young" amine granules are particularly active in a functional aspect have been discussed elsewhere (*e.g. Häggendal* and *Dahlström,* 1972; *Dahlström* and *Häggendal,* 1972, 1973).

d) Mechanisms of Transport

In 1968, Schmitt suggested that microtubules could be involved in fast intra-axonal transport of neuronal organelles (*Schmitt,* 1968). To test this suggestion colchicine (COL) and vinblastine (VIN), two mitotic inhibitors which act by interfering with microtubules, were applied locally to adrenergic neurons. Both substances could interrupt the proximo-distal transport of amine storage granules in a dose dependent manner (*Dahlström,* 1968, 1971 b; *Hökfelt* and *Dahlström,* 1971). These observations gave support to the idea that microtubules were the structural backbone of fast intra-axonal transport of amine granules, but the results could be due to other effects of the substances (*e.g. Wilson et al.,* 1970). Further evidence for the participation of microtubules in amine granule transport were, however, published recently (*Banks et al.,* 1971 a). These authors found that with increasing concentrations of COL in the medium, the transport of

NA and dense core vesicles was decreasing. In parallel with this decreased transport the number of microtubules per axon was also decreased. *Banks et al.* (1971 b) could also demonstrate a morphological relationship between dense core vesicles and microtubules which may be of functional significance.

The force for proximo-distal transport is clearly generated locally in the axon, since transport continues in isolated nerve segments (*Dahlström,* 1967 c; *Banks et al.,* 1969). The mechanism appears to require a continuous supply of some high energy compound, like ATP or GTP, resulting from oxidative metabolism in mitochondria (cf. *Banks, Mayor* and *Mraz,* 1973).

For further discussion on intra-axonal transport in adrenergic nerves, see review articles by *Dahlström* (1971 a), *Banks* and *Mayor* (1972), *Dahlström et al.* (1974 c).

II. Cholinergic Neurons

For technical reasons the cholinergic neurons have so far been studied much less than the adrenergic ones with respect to transport of transmitter and transmitter metabolizing enzymes. There exists as yet no histochemical method for visualization of ACh, and this substance is more difficult to handle in subcellular fractionation studies than is NA. ACh has mainly been quantified by bioassay methods, because other methods have, until recently, been insensitive and unreliable. The metabolizing enzymes, cholineacetyltransferase (CAT) and acetylcholinesterase (AChE) can, however, be studied histochemically (*e.g. Burt* and *Silver,* 1973; *Shute* and *Lewis,* 1966), and a transport intra-axonally of AChE was earlier studied with histochemical and biochemical methods (*e.g. Lubińska et al.,* 1964; *Lubińska,* 1964).

In 1967, *Evans* and *Saunders* described the increase of ACh in regenerating cholinergic motor nerves several days after axotomy. In such "long-term" experiments it is, however, difficult to separate between the effects of degeneration, regeneration and interrupted axonal transport. During the last years it has been demonstrated that ACh accumulates rapidly proximal to a crush or cut in rat sciatic and cervical preganglionic nerve (*Häggendal et al.,* 1971, 1973; *Dahlström et al.,* 1974 a).

a) Sciatic Nerve

The nerves were usually crushed with a thin silk suture *a.m. Lubińska* (1959). At different times after operation 5 mm segments relative to the crush were dissected out and assayed for ACh content, using the guinea pig ileum preparation according to *Blaber* and *Cuthbert* (1961). The ACh content in the 5 mm segments on either side of the crush increased somewhat soon (1—2 min) after crushing. This was probably due to local synthesis of ACh due to the nerve trauma *per se* (cf. *Feldberg,* 1943). Thereafter ACh continued to increase in the part proximal but not distal to the crush (Fig. 2). By 12 hours the level in this segment had increased up to almost 3 times control. In the part just distal to the crush a small further increase up to 160 % of control was observed at 6 hours. In the more distal parts of the nerve (segments C—F in Fig. 2) a decline in ACh levels of about 20 % was observed 6—12 hours after crushing. In a separate experiment where the "transportable fraction" of ACh (see chapter I, a) was studied, the decrease (and thus probably the transportable fraction) was 25 % of control levels at 6 hours after crushing.

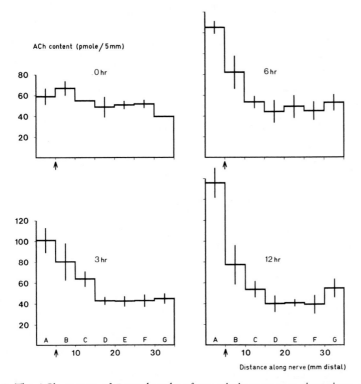

Fig. 2. The ACh content of 5 mm lengths of rat sciatic nerv at various times after crushing. Zero hours = nerves crushed with thread and dissected out immediately. Vertical bars are ± S.E.M. Arrows indicate the site of lesion. n = 3—4, except where indicated. Letters identify nerve segments (from *Dahlström et al.,* 1974 b)

In Fig. 3 the results from Fig. 2 are summarized. Curve I in Fig. 3 a shows the total increase in ACh in segment A just proximal to the crush. The white bars in Fig. 3 b indicate the ACh increase in segment B just distal to the crush. This increase is probably due (at least to a large extent) to local ACh synthesis caused by the crush. Since a similar synthesis is likely to have taken place also in segment A, the increase above control in segment B was subtracted from the total amounts in segment A. In this way curve II in Fig. 3 a was produced. This corrected curve is more likely to show the net accumulation of ACh due to transport. On the basis of curve II and

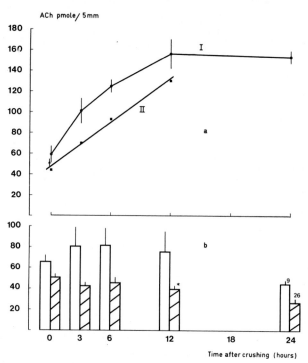

Fig. 3. a) Change in ACh content in 5 mm *proximal* to crush (part A) 0 to 24 hours after operation (I). ●—● original experimental values. ▲ uncrushed nerve. (II) ■—■ values corrected for local synthesis by subtracting "extra" ACh accumulated in part B (Fig. 3 b) above level in uncrushed nerve, from the original experimental values of part A. Vertical bars are ± S.E.M. n = 8—9, except for 0 hour crushed nerves (n = 5).

b) Change in ACh content *distal* to crush 0—24 hours after operation. Open rectangles (B), 5 mm immediately distal to crush. Hatched rectangles (D—F), 10—25 mm distal to crush, cf. Fig. 2. Vertical bars are ± S.E.M. For B, n = 3—4, and for D—F, n = 9—12, except where indicated.

   * Significantly less than 0 hour (p < .05) (from *Dahlström et al.*, 1974 b)

taking into account that only 20—25 % of the ACh is transportable or mobile in this nerve, a rate of transport of the mobile ACh fraction of 4—5 mm/hour has been calculated (*Dahlström et al.*, 1974 a).

It may be argued that the large increase in ACh content in segment A could be due to an accumulation in this segment of CAT which could induce local ACh synthesis. However, in experiments where the accumulation of CAT activity in this segment was determined, we found (*Saunders et al.*, 1973) that this enzyme accumulated very slowly in contrast to ACh (Fig. 4). The results indicated that CAT was transported with the slow flow (2—3 mm/day) (see also *Wooten* and *Coyle*, 1973), but it is possible that a small fraction of CAT in this nerve (less than 5 %) could be rapidly transportable. Provided that the supply of choline and acetyl-Co-A is sufficient, the small possible increase in CAT activity could conceivably induce an increased ACh synthesis in segment A. However, AChE is also present in this segment and would probably metabolize any ACh which was not protected from its action. Therefore, it is likely that the increased levels of ACh in segment A (after correction for crush-induced synthesis according to curve II in Fig. 3 a) is an indication of an increased number of ACh storage units.

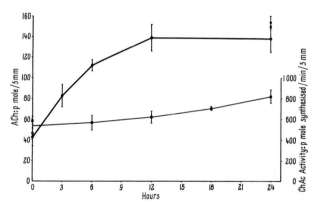

Fig. 4. The content of ACh (•—•) and CAT activity (■—■) in the 5 mm segment of rat sciatic nerve proximal to a crush (part A) at different times after operation. Bars indicate S.E.M., n = 8—9 in the top curve (ACh) and n = 8 in the lower figure (ChAc = CAT) except at 6 hours where n = 4. The CAT curve is taken from *Saunders et al.*, 1973

b) Cervical Preganglionic Nerve

In this nerve, a much smaller increase above control levels of ACh was found on the proximal side of a cut. At 12 hours after cutting the nerve the ACh content proximal to the cut was only 40 % above normal. However, in this nerve only 5 % of the total ACh in control nerves was transportable. When this small transportable fraction was taken into account, the transport rate of this fraction was calculated to about 4 mm/hour, *i.e.* a rapid transport of the same rate as that found to occur in the sciatic nerve (*Häggendal et al.*, 1973).

c) Mechanism of Transport

COL and VIN have been applied to the sciatic nerve at a proximal level by subepineural injections. Two hours after this injection a distal crush was performed and the nerve segments above and around the site of injection, and the part just above the crush, were dissected out 6 hours later (Fig. 5). With increasing concentrations of COL and VIN the ACh content *rose* in the nerve parts above and around the injection site. In parallel with this rise, the amount of ACh which accumulated in the segment just above the crush *decreased*. This indicates that COL and VIN were able to arrest the transport of ACh. With the highest concentrations of COL and VIN the ACh level in the 5 mm segment just above the crush was about

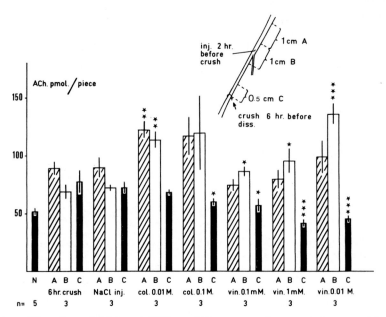

Fig. 5. The effect of COL and VIN on ACh-transport in rat sciatic nerve. Mean ± S.E.M. are given, n = number of observations. The ordinate shows the amount of ACh in pmol per 5 or 10 mm segments after different treatments (indicated along the abscissa). Stars indicate levels of significance, tested against control nerves (from *Dahlström et al.*, 1974 b)

15—20 % *lower than control.* This fraction may correspond to the transportable fraction of ACh, which, following the block of axonal ACh transport by COL and VIN, probably has been conveyed into the more distal parts of the nerve (*Dahlström et al.,* 1974 b).

The results may indicate that axonal microtubules are involved in ACh transport. *Järlfors* and *Smith* (1970) found, in *lamprey* motor nerves, a close morphological association between synaptic vesicles and microtubules. However, the interpretation of the VIN-COL results must be done with great care, since both drugs, in addition to their microtubule-destroying ability, may also have other chemical effects in the axon (*e.g. Wilson et al.,* 1970). It has recently been reported, however, that rapid transport of labelled proteins in rabbit vagus nerve was irreversibly interrupted by COL, and that this block of rapid axonal transport was associated with a progressive loss of axonal microtubules (*Fink et al.,* 1973). This may furnish additional support for the theory that axonal microtubules may be involved in fast intra-axonal transport also in cholinergic neurons.

d) Transport Organelle for ACh

As discussed above the rapidly accumulating ACh is probably stored in a way which protects it from the degradation by AChE. This may indicate a binding to or within a subcellular compartment. Also, the fact that a certain fraction of the axonal ACh is *rapidly* transportable, indicates its association with an organelle (cf. *Skrangiel-Kramska et al.,* 1969). The "non-mobile" rest of the ACh in axons may by contrast be located freely in the axoplasm. This is consistent with earlier observations, suggesting that most axonal ACh is free in the axoplasm (*Carlini* and *Green,* 1963; *Evans* and *Saunders,* 1973).

The suggested transport organelle for ACh is not yet known. However, in subcellular fractionation studies on cat sciatic nerves crushed 18 hours prior to dissection, a large part of the accumulated ACh appeared to be related to particles with a higher density than ordinary synaptic vesicles (*Dahlström et al.,* 1974 b; *Heilbronn et al.,* unpublished). Also, E.M. studies of crushed cholinergic (motor and preganglionic sympathetic) nerves have revealed the accumulation of large dense core vesicles proximal, but not distal, to the crush (*Woog* and *Bennett,* 1972). Thus, the transport organelle for ACh may possibly be similar to the NA transport organelle, *i.e.* a rather large vesicle with a protein matrix. This organelle could conceivably be a precursor of the smaller synaptic vesicles in cholinergic nerve terminals. However, much work needs to be done to elucidate this intriguing question.

Acknowledgements

Supported by the Swedish Medical Research Council (grants Nos. 14X-2207 and 04P-4173), the Magnus Bergwall Foundation, Stockholm, and the Faculty of Medicine, University of Göteborg, Sweden.

References

Andén, N.-E., A. Carlsson, and *J. Häggendal:* Adrenergic mechanisms. Ann. Rev. Pharmacol. *9,* 119—134 (1969).

Banks, P., D. Mangnall, and *D. Mayor:* The redistribution of cytochrome oxidase, noradrenaline and adenosine triphosphate in adrenergic nerves constricted at two points. J. Physiol. *200,* 745—762 (1969).

Banks, P., and *D. Mayor:* Intra-axonal transport in noradrenergic neurons in the sympathetic neurons system. Biochem. Soc. Symp. *36,* 133—149 (1972).

Banks, P., D. Mayor, M. Mitchell, and *D. Tomlinson:* Studies on the translocation of noradrenaline-containing vesicles in post-ganglionic sympathetic neurones *in vitro.* Inhibition of movement by colchicine and vinblastine and evidence for the involvement of axonal microtubules. J. Physiol. *216,* 625—639 (1971 a).

Banks, P., D. Mayor, and *R. D. Tomlinson:* Further evidence for the involvement of microtubules in the intra-axonal movement of noradrenaline storage granules. J. Physiol. (London) *219,* 755—761 (1971 b).

Banks, P., D. Mayor, and *P. Mraz:* Metabolic aspects of the synthesis and intra-axonal transport of noradrenaline storage vesicles. J. Physiol. (London) *229,* 383—394 (1973).

Bisby, M. A., and *M. Fillenz:* The storage of noradrenaline in sympathetic nerve terminals. J. Physiol. (London) *215,* 163—179 (1971).

Blaber, L. C., and *A. W. Cuthbert:* A sensitive method for the assay of acetylcholine. J. Pharm. Pharmac. *13,* 445—446 (1961).

Brimijoin, S.: Transport and turnover of dopamine-β-hydroxylase in sympathetic nerves of the rat. J. Neurochem. *19,* 2183—2193 (1972).

Burt, A., and *A. Silver:* Histochemistry of choline acetyltransferase: a critical analysis. In: Dynamic Aspects of the Synapse, Symposium held in Boldern, Zürich, April, 1973. Brain Research (in press).

Carlini, E. A., and *J. P. Green:* Acetylcholine activity in the sciatic nerve. Biochem. Pharmac. *12,* 1367—1376 (1963).

Corrodi, H., and *G. Johnsson:* The formaldehyde fluorescence method for the histochemical demonstration of biogenic amines. A review on the methodology. J. Histochem. Cytochem. *15,* 65—78 (1967).

Dahlström, A.: Observations on the accumulation of noradrenaline in the proximal and distal parts of peripheral adrenergic nerves after compression. J. Anat. (London) *99,* 677—689 (1965).

Dahlström, A.: The intraneuronal distribution of noradrenaline and the transport and life-span of amine storage granules in the sympathetic adrenergic neuron. Naunyn-Schmiedebergs Arch. Pharmak. exp. Path. *257*, 93—115 (1967 a).

Dahlström, A.: The effect of reserpine and tetrabenazine on the accumulation of noradrenaline in the rat sciatic nerve after ligation. Acta physiol. scand. *69*, 167—179 (1967 a).

Dahlström, A.: The transport of noradrenaline between two simultaneously performed ligations of the sciatic nerves of rat and cat. Acta physiol. scand. *69*, 158—166 (1967 b).

Dahlström, A.: Effect of colchicine on transport of amine storage granules in sympathetic nerves of rat. Europ. J. Pharmacol. *5*, 111—113 (1968).

Dahlström, A.: Axoplasmic transport (with particular respect to adrenergic neurons). Phil. Trans. Roy. Soc. (London) *B 261*, 325—358 (1971 a).

Dahlström, A.: Effects of vinblastine and colchicine on monoamine-containing neurons of the rat, with special regard to the axoplasmic transport of amine granules. Acta neuropath. (Berl.) Suppl. V, 226—237 (1971 b).

Dahlström, A., C. A. N. Evans, J. Häggendal, P.-O. Heiwall, and *N. R. Saunders:* Rapid transport of acetylcholine in rat sciatic nerve proximal and distal to a lesion. J. Neural Transm. *35*, 1—11 (1974 a).

Dahlström, A., and *J. Häggendal:* Studies on the transport and life-span of amine storage granules in a peripheral adrenergic neuron system. Acta physiol. scand. *67*, 278—288 (1966).

Dahlström, A., and *J. Häggendal:* On the possible relation between different pools of adrenergic transmitter and heterogeneity or amine storage granules in nerve terminals. Acta Physiol. Pol. *23*, Suppl. 4, 537—549 (1972).

Dahlström, A., and *J. Häggendal:* Intra-axonal transport of young amine granules: implications for nerve-terminal function. In: Metabolic Regulation and Functional Activity in the Central Nervous System *(Genazzani* and *Herken,* eds.), pp. 94—103. Berlin-Heidelberg-New York: Springer. 1973.

Dahlström, A., J. Häggendal, E. Heilbronn, P.-O. Heiwall, and *N. R. Saunders:* Proximo-distal transport of acetylcholine in peripheral cholinergic neurons. In: Dynamics of Degeneration and Growth in Neurons *(Fuxe, K.,* and *Y. Zotterman,* eds.), pp. 275—290. Stockholm: Pergamon Press. 1974 b.

Dahlström, A., J. Häggendal, P.-O. Heiwall, P.-A. Larsson, and *N. R. Saunders:* Intra-axonal transport of neurotransmitters in mammalian neurons. In: Transport at the Cellular Level. Soc. Exp. Biol. Symp. 28. Cambridge: Univ. Press. 1974 c, in press.

Dahlström, A., J. Häggendal, and *P.-A. Larsson:* On the noradrenaline loading in axonal amine storage granules in rat crushed sciatic nerves. Acta physiol. scand. (in press, 1975).

Dairman, W., L. Geffen, and *M. Marchelle:* Axoplasmic transport of aromatic L-amino acid decarboxylase (EC 4.1.1.26) and dopamine-β-hydroxylase (EC 1.14.2.1) in rat sciatic nerve. J. Neurochem. *20*, 1617 to 1623 (1973).

De Potter, W. P.: Noradrenaline storage particles in splenic nerve. Phil. Trans Roy. Soc. London B *261,* 313—317 (1971).

Euler, U. S. v., and *N.-A. Hillarp:* Evidence for the presence of noradrenaline in submicroscopic structure of adrenergic axons. Nature (London) *177,* 44—45 (1956).

Evans, C. A. N., and *N. R. Saunders:* The distribution of acetylcholine in normal and in regenerating nerves. J. Physiol. *192,* 79—92 (1967).

Falck, B., and *C. Owman:* A detailed methodological description of the fluorescence method for the cellular demonstration of biogenic monoamines. Acta univers. lund., sect. II, 7, 1 (1965).

Feldberg, W.: Synthesis of acetylcholine in sympathetic and cholinergic nerve. J. Physiol. *101,* 432—445 (1943).

Geffen, L. B., and *A. Ostberg:* Distribution of granular vesicles in normal and constricted sympathetic neurones. J. Physiol. *204,* 583—592 (1969).

Häggendal, J.: An improved method for the fluorimetric determination of small amounts of adrenaline and noradrenaline in plasma and tissue. Acta physiol. scand. *59,* 242—254 (1963).

Häggendal, J., and *A. Dahlström:* The recovery of the capacity for uptake-retention of ^{3}H-noradrenaline in rat adrenergic nerves after reserpine. J. Pharm. Pharmacol. *24,* 565—574 (1972).

Häggendal, J., A. Dahlström, S. Bareggi, and *P.-A. Larsson:* Importance of axoplasmic transport for transmitter levels in nerve terminals. In: Dynamics of Degeneration and Growth in Neurons (*Fuxe, K.,* and *Y. Zotterman,* eds.), pp. 257—268. New York: Pergamon Press. 1974.

Häggendal, J., A. Dahlström, and *P.-A. Larsson:* Rapid transport of noradrenaline in adrenergic axons of rat sciatic nerve distal to a crush. Acta physiol. scand. (in press, 1975).

Häggendal, J., A. Dahlström, and *N. Saunders:* Axonal transport and acetylcholine in rat preganglionic neurons. Brain Res. In press (1973).

Häggendal, J., N. R. Saunders, and *A. Dahlström:* Rapid accumulation of acetylcholine in nerve above a crush. J. Pharm. Pharmacol. *23,* 552 to 555 (1971).

Hökfelt, T.: Distribution of noradrenaline storage particles in peripheral adrenergic neurons as revealed by electron microscopy. Acta physiol. scand. *76,* 427—440 (1969).

Hökfelt, T., and *A. Dahlström:* Effects of two mitosis inhibitors (colchicine and vinblastine) on the distribution and axonal transport of noradrenaline storage particles. Studied by fluorescence and electron microscopy. Z. Zellforsch. *119,* 460—482 (1971).

Iversen, L. L.: The Uptake and Storage of Noradrenaline in Sympathetic Adrenergic Nerves. London: Cambridge University Press. 1967.

Järlfors, U., and *D. S. Smith:* Association between synaptic vesicles and neurotubules. Nature (London) *224,* 710—711 (1969).

Kapeller, K., and *D. Mayor:* Ultrastructural changes proximal to a constriction in sympathetic axons during first 24 hours after operation. J. Anat. (London) *100,* 439—441 (1966).

Katz, B.: Nerve, Muscle and Synapse. New York: McGraw-Hill. 1966.

Laduron, P., and *F. Belpaire:* Transport of noradrenaline and dopamine-β-hydroxylase in sympathetic nerves. Life Sci. *7,* 1—7 (1968).

Livett, B. G., L. B. Geffen, and *R. A. Rush:* Immunohistochemical evidence for the transport of dopamine-β-hydroxylase and a catecholamine binding protein in sympathetic nerves. Biochem. Pharmacol. *18,* 923 to 924 (1968).

Lubińska, L.: Region of transition between preserved and regenerating parts of myelinated nerve fibers. J. Comp. Neurol. *113,* 315—335 (1959).

Lubińska, L.: Axoplasmic streaming in regenerating and in normal nerve fibres. In: Mechanisms of Neural Regeneration (*Schadé, J. P.,* and *M. Singer,* eds.). (Progr. Brain Res. 13, pp. 1—71.) Amsterdam: Elsevier. 1964.

Lubińska, L., S. Niemierko, B. Oderfeld-Nowak, and *L. Szwarc:* Behaviour of acetylcholinesterase in isolated nerve segments. J. Neurochem. *11,* 493—503 (1964).

Ochs, S., and *N. Ranish:* Characteristics of the fast transport systems in mammalian nerve fibers. J. Neurobiol. *1,* 247—261 (1969).

Roth, R. H., L. Stjärne, F. E. Bloom, and *N. J. Giarman:* Light and heavy norepinephrine storage particles in the rat heart and in bovine splenic nerve. J. Pharmacol. exp. Ther. *162,* 203—212 (1968).

Saunders, N. R., K. Dziegelewska, J. Häggendal, and *A. Dahlström:* Slow accumulation of choline acetyltransferase in crushed sciatic nerves of the rat. J. Neurobiol. *4,* 95—103 (1973).

Schmitt, F. O.: The molecular biology of neural fibrous proteins. Neurosci. Res. Progr. Bull. *6,* 119—144 (1968).

Shute, C. C. D., and *P. R. Lewis:* Electron microscopy of cholinergic terminals and acetylcholinesterase-containing neurones in the hippocampal formation of the rat. Z. Zellforsch. *69,* 334—343 (1966).

Skrangiel-Kramska, J., S. Niemierko, and *L. Lubińska:* Comparison of the behaviour of a soluble and a membrane-bound enzyme in transected peripheral nerves. J. Neurochem. *16,* 921—926 (1969).

Wilson, L., J. Bryan, A. Ruby, and *D. Mazia:* Precipitation of proteins by vinblastine and calcium ions. Proc. Nat. Acad. Sci. *66,* 807—814 (1970).

Woog, R. H., and *M. R. Bennett:* Vesicle types in injured nerve fibres. Proc. Austr. Physiol. Pharmacol. Soc. *3,* 89—90 (1972).

Wooten, G. F., and *J. T. Coyle:* Axonal transport of catecholamine synthesizing and metabolizing enzymes. J. Neurochem. *20,* 1361—1371 (1973).

Author's address: Dr. *Annica Dahlström,* Institute of Neurobiology, University of Göteborg, Medicinareg. 5, Fack, S-400 33 Göteborg 33, Sweden.

Discussion

Zenker: Does the effect of colchicin on neurotubules really explain the involvement of neurotubules in the mechanism of axonal flow? From the literature I know various effects of colchicin on the appearance of neurotubules. So on one hand it was shown that neurotubules desintegrate upon colchicin whereas a great augmentation of neurotubules after colchicin was shown by other authors.

Dahlström: I agree with Dr. Zenker that the various results are sometimes puzzling. One would need to go through the studies carefully and range the results according to dose, mode of application, time after application and temperature during fixation. I think many contradictory results may be explained if this was done. It is for instance possible that reversion of the colchicine effect may include, in some cases, a compensatory, increased polymerization of soluble microtubule subunits.—Naturally, we must also keep in mind that *perhaps* the microtubules are not involved in particulate transport in the way we imagine.

Pilgrim: In this connection, I should like to mention an observation by Dr. Flament-Durand, Brussels, which she reported at the recent Neurosecretion meeting at London. She investigated the influence of colchicine on the transport of neurosecretory granules from the paraventricular nucleus. There was a heavy accumulation of granules in the nerve cell bodies as well as in axon swellings indicating interference of the drugs with axoplasmic transport, but no change in the number of microtubules.

Dahlström: Norström et al. (1970) also reported an inhibited axonal transport of neurosecretory granules in rat supraoptic neurons with no marked decrease in the average number of tubules. However, their observations were made on rats given colchicine for 2 days prior to sacrifice when reactions to the treatment have no doubt occurred. This, to my mind, is too long. Also, their pictures show neurons which have reacted very differently. Some axons have many tubules, but few arrested granules. Others have a lot of granules, but few visible tubules. I am not familiar with Dr. Flament-Durand's report, but I suggest that short time intervals should be used for studies on colchicine effects.

Whittaker: May I comment on your attempt to isolate acetylcholine-containing vesicles from nerve axons. In my experience, homogenizing axons is like homogenizing rubber bands. Might not the dense acetylcholine-containing particulate material simply consist of large fragments of incompletly homogenized nerve trunk? Has any morphology been done on these fractions?

I was also puzzled by the appreciable amount of free acetylcholine recovered in the top fraction of the density gradient. In view of the presence of cholinesterase in the preparation, how do you account for the survival of free acetylcholine?

Dahlström: The ACh in the heavy fractions could, of course, be present in occluded axoplasm. However, choline-acetyltransferase activity in these fractions was very low or non-detectable. Most of this enzyme is soluble, and if axoplasm was occluded in the heavy fractions one ought to have discovered this enzyme in the fractions. AChE, however, was present in large amounts. The ACh must therefore be present in an esterase-protected form.

The large amounts of ACh in the supernatant despite the absence of eserine are puzzling. Dr. Heilbronn speculates if possibly an AChE-inhibitor might be present in the system.

Kobinger: Do you have any evidences or hypothesis about an information-system from the peripheral part of the nerve to the nerve proximal of the crush, which may induce the increase in transmitter synthesis?

Dahlström: One possible way would be by indirect means—a reflectory mechanism causing an increased impulse activity in the preganglionic neuron which in turn induces an enhanced synthesis of proteins, including transmitter storage granules. If the preganglionic impulse activity is blocked the increase in NA granule production and in DBH production (*Mueller et al.,* 1970) is prevented. This finding thus supports the reflex mechanism. Of course, other mechanisms may operate, *e.g.* the suggested retrograde transport of protein molecules.

Sotelo: 1. A good evidence in favour of the active role of microtubules in the fast axonal flow of synaptic vesicles will be to find a specific binding affinity between tubuli and the vesicular membrane protein. To my knowledge such affinity has not been proven.—Thus, after you, what is the mechanism for such a process?

2. In sympathetic axonal varicosities of the pineal gland it seems that there are two different substances, both of them being good candidates for neurotransmitters: noradrenaline and serotonin. According to histochemical work of Jaim-Echeverry (reference) it looks like serotonin is in LGV and noradrenaline in SGV. How do these results fit with your hypothesis that LGV are only younger storage granules than SGV?

Dahlström: 1. It would be nice, of course, to find such an affinity. However, if one considers the suggested mechanism for transfer (a "sliding" mechanism where specific sites are alternatingly attractive and repulsive) it is not entirely expected to find only attractive forces operating in a homogenized system. Also, *e.g.* Ochs has suggested another mechanism where a specific vehicle molecule should act between vesicles and tubules. Neither in this case one would expect to find a specific binding between vesicles and tubulin.

2. The results that 5-HT may be mostly localized in LGV fit very well in with our hypothesis that LGV may represent young amine granules. The 5-HT in the pineal nerve terminals is likely to have been produced in the pinealocytes and taken up into the nerve by the membrane pump mechanism. This 5-HT is thus "exogeneous" to the nerve. One important capacity of young amine granules was, according to our results, the capacity to take up and store "exogeneous" amine. The result by Jaim-Etcheverry that LGV stores the 5-HT thus fits very well with our hypothesis that LGV may correspond to young amine granules.

Fleischhauer: I am sorry, I did not quite understand the last point of the discussion. Has it been shown beyond doubt that more than one transmitter substance may be present in one and the same terminal? Is one cell capable of producing two different transmitters?

Dahlström: This is a particular case; the noradrenergic nerve terminals in the corpus pineale. These terminals belong to neurons in the superior cervical ganglion. Their "endogeneous" transmitter is NA, but since they, in the pineal gland, are located in a very 5-HT-rich environment, the terminals take up 5-HT, like they take up exogeneous NA. The 5-HT is thus presumably *not formed* within the terminals. To my knowledge, it is not demonstrated that 5-HT, in addition to NA, acts as a transmitter. This, however, would be rather plausible.

Journal of Neural Transmission, Suppl. XII, 115—126 (1975)

Regulation Mechanisms of Central and Peripheral Sympathetic Neurons

G. Hertting and B. Peskar

Dept. of Pharmacology, University of Freiburg i. Br., Federal Republic of Germany

With 6 Figures

For many years, the main interest in mechanisms regulating the activity of peripheral sympathetic neurons has been focussed on the structures which maintain the homoeostasis of blood circulation or adapt the circulation to the momentary needs. The mechanisms for this type of regulation receive their information through pressor-, chemo- or volume receptors. This information is then integrated in central structures, and the efferent impulses adapt the periphery to the specific situation. The peripheral activity of the sympathetic nervous system, however, can in addition be modified by local processes that adjust their activity and function to local needs which had escaped being monitored because they were too specialized to be regulated by an overall adjustment, or even apparently controversial to it. In the last decade the knowledge about these local mechanisms has increased, and some of them will be briefly discussed in this paper.

A nerve impulse that reaches the sympathetic periphery, liberates the sympathetic transmitter noradrenaline (NA), which leaves the nerve, passes across the synaptic cleft and stimulates the receptors on the target cell. The stimulation frequency and the duration of the stimulus trains is regulated by the central projections of the sympathetic nervous system. However, besides this, a number of mechanisms is available to regulate the amount of transmitter released per stimulus, to adjust the concentration of transmitter that reaches the receptor sites and to modify the duration of the action of the transmitter as well as the response of the effector cells. These mechanisms must be controlled locally to serve the isolated local necessities.

Factors Influencing the Neuronal Uptake Process of Noradrenaline

The neuronal uptake process is the most efficient mechanism regulating the concentration of the free and active local hormone NA in the synaptic cleft and therefore at the receptor sites. It is an active transport mechanism, temperature dependent, that can be inhibited by metabolic inhibitors (*Kirpekar* and *Wakade*, 1968; *Wakade* and *Furchgott*, 1968). This "membrane pump" is possibly linked to the sodium pump of the axonal membrane (*Bogdanski* and *Brodie*, 1969) since it has an absolute requirement for sodium ions (*Iversen* and *Kravitz*, 1966; *Bogdanski* and *Brodie*, 1966; *Kirpekar* and *Wakade*, 1968; *Horst, Kopin* and *Ramey*, 1968) and can be blocked by ouabain (*Dengler, Spiegel* and *Titus*, 1961) as well as high potassium concentration (*Bogdanski* and *Brodie*, 1969). This neuronal uptake process "uptake I" (*Iversen*, 1963) is further characterized by high affinity, but low capacity for NA. At physiological stimulus frequencies a considerable amount of NA released by stimuli is reaccumulated into the postganglionic sympathetic fiber (*Brown*, 1965). At some organs, *e.g.* the spleen, the termination of the sympathetic activity depends primarily on this neuronal uptake process, the metabolism of NA by monoamine oxydase (MAO) or catechol-O-methyl transferase (COMT) being negligeable (*Hertting* and *Suko*, 1972).

It is obvious that factors interfering with the uptake I process will increase the NA concentration at the receptor sites and maintain it for a longer period of time, producing an increased and prolonged effect. In organs with very dense sympathetic innervation, as in the iris or the vasculature, the potentiation of sympathetic activity by this mechanism will of course be of greater importance than in regions with sparse sympathetic innervation, as for example in thick-walled blood vessels. The first figure shows a schema of the events at the sympathetic periphery: The nerve impulse liberates the NA stored in the intraneuronal granules, the NA enters the synaptic cleft, some of it reaches the receptor sites at the smooth muscle cells producing a contraction. A fraction of the released NA is taken up again into the sympathetic neuron before it can act at the receptor.

A number of drugs have a blocking action on this neuronal uptake process, as the tricyclic antidepressives, neuroleptics and others. The model drug in this respect is cocaine. Under cocaine, the uptake mechanism is blocked and the tissues become supersensitive to NA (*Whitby, Hertting* and *Axelrod*, 1960; *Muscholl*, 1960). Similarly, after cocaine pretreatment the effect of sympathetic nerve stimulation is increased due to the increase in transmitter release, as has been

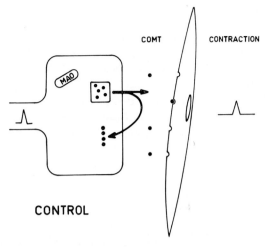

Fig. 1. The figure pictures the events at the sympathetic nerve terminal. The incoming nerve impulse liberates the neurotransmitter noradrenaline (NA). Some of the NA released produces a contraction of a smooth muscle cell by interacting with the α-receptors. A large share of the NA is transported back into the nerve terminal.

MAO = monoamine oxidase; COMT = catechol-O-methyl-transferase; • = NA quantum; ☐ = storage granule

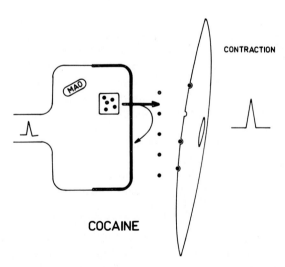

Fig. 2. Symbols see Fig. 1. Cocaine, or compounds with similar actions, block neuronal re-uptake. More transmitter remains available for the receptor sites, producing an increased response

shown for the spleen (*Thoenen, Huerlimann* and *Haefely,* 1964) or the heart (*Hukovic* and *Muscholl,* 1962). Likewise, potentiation of the endogenous sympathetic activity occurs under cocaine or similar compounds: Administration to the eye produces dilatation of the pupil by increasing the tonus of the sympathetically innervated dilatator pupillae.

Fig. 2 pictures this action of cocaine or cocaine-like drugs: Again, the nerve impulse liberates the transmitter, but the re-uptake at the nerve membrane is blocked, a higher concentration of the free and active hormone persists for a longer period of time in the vicinity of the receptors, producing an increased response.

One must consider, however, that the systemic administration of drugs with cocaine-like action will mimic an increase of sympathetic activity, which will lead to a re-adjustment of the central sympathetic outflow.

It is conceivable that changes in the intracellular and extracellular electrolyte pattern, acid-base equilibrium or metabolic requirements can influence the NA-uptake process and contribute to changes of sympathetic activity.

Neuronal α-Receptor Feed-Back Mechanism

It was recognized as early as 1957 by *Brown* and *Gillespie* that following the blockade of α-receptors larger amounts of NA were liberated after stimulation of sympathetic nerves. The "overflow" of NA obtained in such preparations after α-receptor-blockade was first thought to be equivalent to the amount of NA that was prevented from reaching the α-receptors in the periphery. Moreover, in addition to this property some of the α-blocking drugs also blocked neuronal uptake. It has therefore been proposed that the increased transmitter discharge following α-receptor blockade is due to a decreased neuronal re-uptake and decreased tissue binding, followed by metabolic degradation. Several groups, working on this problem, however, have come to the conclusion that it is rather the amount of transmitter discharged by the nerve stimulation which can be increased by α-receptor blocking drugs than the inactivation of the transmitter after being released (*Häggendal,* 1970; *Farnebo* and *Hamberg,* 1971; *Farnebo* and *Malmfors,* 1971; *Langer, Adler, Enero* and *Stefano,* 1971; *Potter, Chubb, Put* and *Schaepdryver,* 1971; *Starke, Montel* and *Wagner,* 1971; *Wennmalm,* 1971 a, b; *Starke,* 1972 a, b). The following hypothesis evolved from their experimental data: Inhibitory α-receptor-like structures are assumed to be located at the sympathetic nerve terminals. These inhibitory α-receptors should modulate the amount of NA released per stimulus. The NA released together with these α-receptors constitutes a negative feed-back loop (*Langer, Adler, Enero* and *Stefano*, 1971; *Starke, Montel* and *Wagner*, 1971; *Starke*, 1972 a, b; *Kirpekar, Wakade, Steinsland, Prat* and *Furchgott*, 1972). Similarly, decreased transmitter release was achieved by various other α-mimetic drugs. The mechanism could also be demonstrated in such organs as the heart which contain only β-receptors at the effector cells (*Starke, Montel* and *Wagner*, 1971; *Starke*, 1972 b). Moreover, transmitter release from brain slices induced by field stimulation was increased following α-receptor-blockade and decreased by adding NA or other α-mimetics to the bath solution (*Starke* and *Montel*, 1973). It can therefore be accepted that adrenergic nerve terminals contain inhibitory α-receptors, which regulate the amount of transmitter released per stimulus. Fig. 3 summarizes this concept: Again, the incoming nerve impulse liberates the transmitter. Some of the transmitter released acts on the receptor sites of the effector cell, some of it exerts an action on the inhibitory α-receptors at the nerve terminal. It is not yet known, whether these receptor sites are located at the mebrane level or if they interfere with the mechanism that makes the granular stores accessible for release.

Introduction of α-blockers (Fig. 4) not only blocks the α-receptors at the effector cell, thereby inhibiting the contraction, but also blocks the inhibitory α-receptors at the nerve terminal, increasing the amount

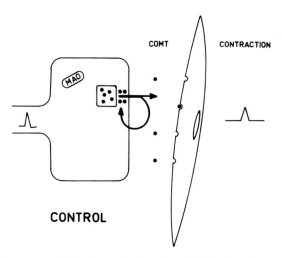

Fig. 3. Symbols see Fig. 1. This figure depicts schematically the actions of α-receptor regulation of NA release. The liberated transmitter modulates transmitter release

of transmitter that is released per stimulus. In organs with mainly β-receptors at the effector cells, as in the heart, α-blockade will produce an increase in the response to nerve stimulation. This mechanism again serves the periphery in adjusting and modulating the sympathetic activity through a local feed-back system. Little is known about the functional value or limits of this mechanism. But it may be assumed that many possible events that change the local metabolic degradation of the released transmitter, such as changes in electrolyte distribution that influence receptor sensitivity or membrane transport mechanisms may induce modifications of the sympathetic activity at this level.

Prostaglandin Mediated Control of Noradrenaline Release

During the last decade the knowledge about another group of compounds, the prostaglandins (Pgs), has increased. The prostaglandins are synthetized in many tissues of the body and serve many different purposes. They are involved not only in the fields of reproduction or inflammation, but may also be involved in various regulatory mechanism. One of their possible roles may be linked to the

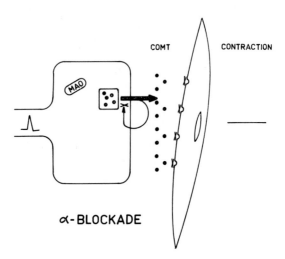

Fig. 4. Symbols see Fig. 1. α-blockade not only prevents the action of the transmitter on the α-receptors of the target cell, but also induces a blockade of the α-adrenergic feed-back inhibition at the neuronal level. Increased amounts of transmitter are released per stimulus

regulation of the activity of sympathetic neurotransmission. It has been shown by *Wennmalm* and *Hedqvist* (1970) and *Hedqvist* (1970) and confirmed by others, that administration of Pgs of the E-series inhibits the release of NA following sympathetic nerve stimulation. This effect was demonstrated in the heart (*Wennmalm* and *Hedqvist*, 1970), seminal vesicle (*Hedqvist*, 1972), in the spleen (*Hedqvist*, 1970), and vas deferens (*Euler* and *Hedqvist*, 1969). During sympathetic stimulation, a fatty acid was released into the perfusate that had similar chemical properties as Pg (*Wennmalm* and *Stjärne*, 1971). The fatty acid-rich perfusate did inhibit, similar to PgE, the transmitter release following stimulation. PgEs were isolated and identified from perfusates of sympathetically stimulated organs (*Gilmore, Vane* and *Wyllie*, 1968; *Davies, Horton* and *Witherington*, 1968). Moreover, inhibition of Pg synthesis by 5, 8, 11, 14- tetraynoic acid enhanced nerve stimulation induced NA-release from the rabbit heart (*Samuelsson* and *Wennmalm*, 1971), cat spleen (*Hedqvist, Stjärne* and *Wennmalm*, 1971), or vas deference (*Hedqvist* and *Euler*, 1972). Similarly, catecholamine excretion in the urine was augmented following inhibition of Pg synthesis by indomethacin (*Junstad* and *Wennmalm*, 1972; *Stjärne*, 1971).

The proposed mechanism of the Pgs modifying sympathetic transmitter release can be summarized as follows (Fig. 5): NA, released upon nerve stimulation from the sympathetic terminals produces an

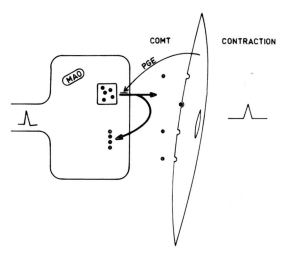

Fig. 5. Symbols see Fig. 1. PgE = Prostaglandin E. PgE is synthetized and released during the contraction of the target cell. It in turn limits stimulus-induced NA release

action on the effector cell. Pgs are synthetized and released following the response from the effector cell and in turn inhibit transmitter release of consecutive stimuli.

Inhibition of Pg synthesis (Fig. 6) relieves the sympathetic nerve terminal of the Pg-induced inhibition, more transmitter is released per stimulus. The efficiency of this feed-back mechanism may vary from tissue to tissue: it seems to be relative unimportant in the brain (*Starke* and *Montel*, 1973) and of varying significance in peripheral organs and tissues. The two inhibitory principles at the sympathetic nerve terminals, namely the α-receptors and the Pgs apparently operate as two basically independent feed-back loops; however they may influence each other. The Pgs that are released during the contraction of smooth muscle cells not only have effects on transmitter release, but can also inhibit or potentiate the contractile response of the smooth muscle cell itself (*Ferreira, Moncada* and *Vane*, 1973; *Peskar* and *Hertting*, 1973). This principle therefore not only modulates the amounts of transmitter released, but also adjusts the response of the effector organ.

It is clear that many additional mechanisms contribute to the regulation of the peripheral functional entity to adapt it to the most efficient functional state. The mechanisms discussed above merely represent those which are better understood.

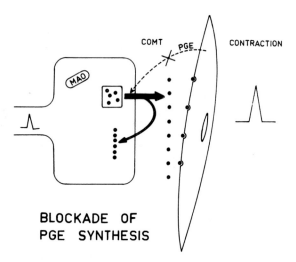

BLOCKADE OF
PGE SYNTHESIS

Fig. 6. Symbols see Figs. 1 and 5. Inhibition of prostaglandin synthesis blocks the Pg-induced feed-back inhibition of NA release, again more NA is released per stimulus and more transmitter reacts with the receptors of the target cell, producing an increased contraction

References

Bogdanski, D. F., and *B. B. Brodie:* Role of sodium and potassium ions in storage of norepinephrine by sympathetic nerve endings. Life Sci. *5*, 1563—1569 (1966).

Bogdanski, D. F., and *B. B. Brodie:* The effects of inorganic ions on the storage and uptake of H^3-norepinephrine by rat heart slices. J. Pharmacol. exp. Ther. *165*, 181—189 (1969).

Brown, G. L., and *J. S. Gillespie:* The output of sympathetic transmitter from the spleen of the cat. J. Physiol. *138*, 81—102 (1957).

Brown, G. L.: The release and fate of the transmitter liberated by adrenergic nerves. Proc. roy. Soc. B 162, 1—19 (1965).

Davies, B. N., E. W. Horton, and *P. G. Witherington:* The occurrence of prostaglandin E_2 in splenic venous blood of the dog following splenic nerve stimulation. Brit. J. Pharmacol. *32*, 127—135 (1968).

Dengler, H. J., H. E. Spiegel, and *E. O. Titus:* Uptake of tritium-labelled norepinephrine in brain and other tissues of the cat in vitro. Science *133*, 1072—1073 (1961).

Euler, U. S. v., and *P. Hedqvist:* Inhibitory action of prostaglandins E_1 and E_2 on the neuromuscular transmission in the guinea pig vas deferens. Acta physiol. scand. *77*, 510—512 (1969).

Farnebo, L. O., and *B. Hamberger:* Drug- induced changes in the release of (^3H)-noradrenaline from field stimulated rat iris. Brit. J. Pharmacol. *43*, 97—106 (1971).

Farnebo, L. O., and *T. Malmfors:* ^3H-noradrenaline release and mechanical response in the field stimulated mouse vas deferens. Acta physiol. scand., Suppl. *371*, 1—18 (1971).

Ferreira, S. M., S. Moncada, and *J. R. Vane:* Some effects of inhibiting endogenous prostaglandin formation on the response of the cat spleen. Brit. J. Pharmacol. *47*, 48—58 (1973).

Gillmore, N., J. R. Vane, and *J. H. Wyllie:* Prostaglandins released by the spleen. Nature *218*, 1135—1140 (1968).

Häggendal, J.: Some further aspects on the release of the adrenergic transmitter. In: New Aspects of Storage and Release Mechanisms of Catecholamines (*Schümann, H. J.*, and *G. Kroneberg*, eds.), pp. 100—111. Berlin-Heidelberg-New York: Springer. 1970.

Hedqvist, P.: Control by prostaglandin E_2 of sympathetic neurotransmission in the spleen. Life Sci. *9*, Pt. 1, 269—278 (1970).

Hedqvist, P., and *U. S. v. Euler:* Prostaglandin controls neuromuscular transmission in guinea pig vas deference. Nature New Biol. *236*, 113—115 (1972).

Hedqvist, P., L. Stjärne, and *Å. Wennmalm:* Facilitation of sympathetic neurotransmission in the cat spleen after inhibition of prostaglandin synthesis. Acta physiol. scand. *83*, 430—432 (1971).

Hedqvist, P.: Prostaglandin induced inhibition of neurotransmission in the isolated guinea pig seminal vesicle. Acta physiol. scand. *84*, 506—511 (1972).

Hertting, G., and *J. Suko:* Influence of neuronal and extraneuronal uptake on disposition, metabolism, and potency of catecholamines. Perspectives in Neuropharmacology, pp. 267—300. Oxford University Press. 1972.

Horst, W. D., I. J. Kopin, and *E. R. Ramey:* Influence of sodium and calcium on norepinephrine uptake by isolated perfused rat hearts. Am. J. Physiol. *215,* 817—822 (1968).

Huković, S., and *E. Muscholl:* Die Noradrenalin-Abgabe aus den isolierten Kaninchenherzen bei sympathischer Nervenreizung und ihre pharmakologische Beeinflussung. Naunyn-Schmiedeberg's Arch. exp. Path. Pharmakol. *244,* 81—96 (1962).

Iversen, L. L.: The uptake of noradrenaline by the isolated perfused rat heart. Brit. J. Pharmacol. *21,* 523—537 (1963).

Iversen, L. L., and *E. A. Kravitz:* Sodium dependence of transmitter uptake at adrenergic nerve terminals. Mol. Pharmacol. *2,* 360—362 (1966).

Junstad, M., and *Å. Wennmalm:* Increased renal excretion of noradrenaline in rats after treatment with prostaglandin synthesis inhibitor indomethacin. Acta physiol. scand. *85,* 573—576 (1972).

Kirpekar, S. M., and *A. R. Wakade:* Factors influencing noradrenaline uptake by the perfused spleen of the cat. J. Physiol. *194,* 609—626 (1968).

Kirpekar, S. M., A. R. Wakade, O. S. Steinsland, J. C. Prat, and *R. F. Furchgott:* Inhibition of the evoked release of norepinephrine (NE) by sympathomimetic amines. Fed. Proc. *31,* 566 Abs. (1972).

Langer, F. Z., E. Adler, M. A. Enero, and *F. J. E. Stefano:* The role of the alpha receptor in regulating noradrenaline overflow by nerve stimulation. Proc. intern. Union Physiol. Sci. *9,* 335 (1971).

Muscholl, E.: Die Hemmung der Noradrenalin-Aufnahme des Gewebes durch Cocain. Arch. exp. Path. Pharmakol. *240,* 8 (1960).

Peskar, B., and *G. Hertting:* Release of prostaglandins from isolated cat spleen by angiotensin and vasopressin. Naunyn-Schmiedeberg's Arch. Pharmacol. *279,* 227—234 (1973).

Potter, W. P. de, I. W. Chubb, A. Put, and *A. F. de Schaepdryver:* Facilitation of the release of noradrenaline and dopamine-β-hydroxylase at low stimulation frequencies by α-blocking agents. Arch. int. Pharmacodyn. *193,* 191—197 (1971).

Samuelsson, B., and *A. Wennmalm:* Increased nerve stimulation induced release of noradrenaline from the rabbit heart after inhibition of prostaglandin synthesis. Acta physiol. scand. *83,* 163—168 (1971).

Starke, K., H. Montel, and *J. Wagner:* Effect of phentolamine on noradrenaline uptake and release. Naunyn-Schmiedeberg's Arch. Pharmak. *271,* 181—192 (1971).

Starke, K.: Alpha sympathomimetic inhibition of adrenergic and cholinergic transmission in the rabbit heart. Naunyn-Schmiedeberg's Arch. Pharmacol. *274,* 18—45 (1972 a).

Starke, K.: Influence of extracellular noradrenaline on the stimulation-evoked secretion of noradrenaline from sympathetic nerves: evidence for an α-receptor-mediated feed-back inhibition of noradrenaline release. Naunyn-Schmiedeberg's Arch. Pharmacol. *275,* 11—23 (1972 b).

Starke, K., and *H. Montel:* Interaction between indomethacin, oxy-metazoline and phentolamine on the release of ^{3}H-noradrenaline from brain slices. J. Pharm. Pharmac. *25,* 758—759 (1973).

Stjärne, L.: Hyperexcretion of catecholamines induced by indomethacin. Acta physiol. scand. *83,* 574—576 (1971).

Thoenen, H., A. Huerlimann, and *W. Haefely:* The effects of sympathetic nerve stimulation on volume, vascular resistance, and norepinephrine output in the isolated perfused spleen of the cat, and its modification by cocaine. J. Pharmacol. exp. Ther. *143,* 57—63 (1964).

Wakade, L. G., and *R. F. Furchgott:* Metabolic requirements for the uptake and storage of norepinephrine by the isolated left atrium of the guinea pig. J. Pharmacol. exp. Ther. *163,* 123—135 (1968).

Wennmalm, Å., and *P. Hedqvist:* Prostaglandin E1 as inhibitor of the sym-pathetic neuroeffector system in the rabbit heart. Life Sci. 9, Pt. 1, 931—937 (1970).

Wennmalm, Å.: Quantitative evaluation of release and reuptake of adren-ergic transmitter in the rabbit heart. Acta physiol. scand. *82,* 532—538 (1971 a).

Wennmalm, Å.: Studies on mechanisms controlling the secretion of neuro-transmitters in the rabbit heart. Acta physiol. scand., Suppl. *365* (1971 b).

Whitby, L. G., G. Hertting, and *J. Axelrod:* Effect of cocaine on the dis-position of noradrenaline labelled with tritium. Nature *187,* 604—605 (1960).

Author's address: Prof. Dr. *G. Hertting* and Dr. *B. Peskar,* Department of Pharmacology, University of Freiburg i. Br., Katharinenstraße 29, D-7800 Freiburg i. Br., Federal Republic of Germany.

Discussion

Kobinger: Possibly the regulation of noradrenalin liberation at adren-ergic nerve endings is not specific. Electrical stimulation of the isolated ileum liberates acetylcholine and this can be inhibited by α-adrenoceptor stimulating substances [*Paton* and *Vizi,* Brit. J. Pharmacol. *35,* 10 (1968)].

Hertting: The conditions in the gut are much more complicated than in other tissues, since this tissue contains ganglia and therefore also intra-neuronal synapses. The inhibitory function of inhibitory sympathetic fibers can be assumed to work at this level, that is at the ganglia. Likewise, catecholamines would act at that level.

Kraupp: There is a certain discrepancy between the above-propounded views on the significance of the prostaglandins in the regulation of the blood flow of various organs and their action spectrum in the intact organism. Are there any means of explaining this discrepancy?

Hertting: The different members of the family of prostaglandins produce diverse effects that may vary from organ to organ and tissue to tissue.

PgEs, for example, stimulate adenyl cyclase [*Berti, F., et al.:* Prostaglandins on cyclic-AMP formation in cerebral cortex of different mammalian species. In: Advances in the Biosciences (*Bergström, S.,* ed.), Vol. 9, pp. 475—480. Oxford: Pergamon Press. 1973]; they may increase blood flow by causing vasodilatation. PgFs, on the other hand, contract blood vessels. It will depend of the pattern of synthesis of the Pgs which effect will become dominant. It will also depend on the amount of the Pg degradating enzymes whether their action will be limited to local structures or produce overall effects.

Journal of Neural Transmission, Suppl. XII, 127—136 (1975)

Trans-Synaptic Regulation of the Synthesis
of Specific Neuronal Proteins

H. Thoenen

Department of Pharmacology, Biocenter of the University, Basel, Switzerland

Introduction

For several decades the main interest in neurobiological research was concentrated on the morphological aspects of neuronal systems and on the electrical manifestations of neuronal activity. The biochemical studies preferentially concerned static-descriptive approaches particularly those referring to macromolecular constituents of neurons. The neurons were thought to represent relatively stable electronic entities, designed to generate, transmit and modulate electrical impulses. This attitude to the interpretation of neuronal function and particularly to interneuronal relationship prompted the comparison of integrated neuronal systems with computers. However, a fundamental difference between a computer and an integrated neuronal system is the capability of the latter to adapt to changes in functional requirements. The capability to adapt to functional requirements is characterized by the term "neuronal plasticity" and involves not only changes in the macromolecular composition of the neurons but even in their morphological features. The basis for such plastic reactions is a relatively rapid turnover of the macromolecules in question and the availability of mechanisms which link the neuronal activity with the regulation of the synthesis of these macromolecules.

Indeed, it has been shown that the neuronal cell body processes a very efficient protein synthesis comparable to that of secretory glands (cf. *Richter*, 1970). Furthermore, the peripheral parts of the neurons, axons and dendrites, are supplied with these macromolecules by an efficient rapid transport system which provides the prerequisite for their highly differentiated functions. The particular capability of neurons for plastic reactions is mainly based on the fact that the response of a neuron to a chemically mediated nerve impulse of

another neuron is not confined to actual effects such as changes in the
ionic permeability of the neuronal membrane, changes in glucose or
O_2 consumption. The response of the effector neurons also involves
longer lasting effects reflected by changes in the regulation of the
synthesis of macromolecular cell constituents.

These aspects of neurobiology have gained increasing interest over
the last few years after it became apparent that such mechanisms are
involved in the regulation of ontogenetic processes (*Black et al.,*
1971 a, b; *Thoenen,* 1972 b) in the long-term adaptation to increased
transmitter utilization (*Thoenen* and *Oesch,* 1973), possibly in the
long-term storage of information in the brain as well as in the
development of tolerance and dependence to centrally acting drugs
which cannot be explained by the induction of drug-metabolizing
enzymes (*Molinoff* and *Axelrod,* 1971; *Thoenen,* 1972 a).

For studying the detailed mechanisms of the regulation of the
synthesis of macromolecules by nerve impulses the highly complex
(both with respect to function and topography) mammalian brain
does not seem to provide favorable experimental conditions. However, the relatively simply organized peripheral sympathetic nervous
system lends itself favorably to such studies.

Specificity of Trans-Synaptic Enzyme Induction in the Peripheral Sympathetic Nervous System

An increased activity of the peripheral sympathetic nervous
system is immediately followed by a rise in the synthesis of catecholamines from tyrosine in the nerve terminals of the adrenergic neuron.
This immediate adaptation to increased utilization of norepinephrine
is not accompanied by an increase in the activity of tyrosine
hydroxylase in enzyme preparations of the corresponding organs
assayed under standardized experimental conditions. The mechanism
of this immediate adaptation to increased transmitter utilization is
not yet established. However, according to the information available
so far changes in endproduct inhibition of tyrosine hydroxylase do
not seem to be responsible for the adaptation, in contrast to the
attenuated norepinephrine synthesis occuring after administration of
monoamine oxidase inhibitors which reflects the competitive inhibition
of the pteridine cofactor of tyrosine hydroxylase by the intra-axonal
accumulation of catecholamines. The aspects of immediate adaptation
of norepinephrine synthesis to increased transmitter utilization
resulting from augmented neuronal activity have recently been
reviewed by *Weiner et al.* (1972).

In addition to this immediate adaptation to augmented transmitter utilization there is another mechanism which comes into play

after prolonged increase in neuronal activity. After the original observation of *Mueller et al.* (1969 a) that the selective destruction of the peripheral sympathetic nerve terminals by 6-hydroxydopamine is followed by a marked increase in the *in vitro* activity of tyrosine hydroxylase of the adrenal medulla (which is not damaged by 6-hydroxydopamine) a great variety of experimental conditions have been found which lead to an increase in the activity of this enzyme both in the adrenal medulla and the terminal adrenergic neurons (cf. *Molinoff* and *Axelrod*, 1971; *Thoenen*, 1972 a). The common denominator of all these experimental conditions is an increased activity of the preganglionic cholinergic nerves. The causal relationship between the increased preganglionic activity and the rise in tyrosine hydroxylase activity in the terminal adrenergic neuron is documented by the finding that transection of the preganglionic cholinergic trunk abolishes the rise in enzyme activity (*Thoenen et al.*, 1969 a).

The increase in tyrosine hydroxylase activity results from an augmented synthesis of new enzyme protein rather than from allosteric changes in enzyme activity by low molecular effectors. This can be deduced from the following observations:

a) The activity of enzyme preparations of control and experimental animals as always additive (*Mueller et al.*, 1969 b).

b) The Km-values for both substrate and cofactor do not differ between enzyme preparations of controls and treated animals (*Mueller et al.*, 1969 b).

c) The increase in enzyme activity can be abolished by inhibitors of protein synthesis acting both at the transcription or translation level (*Mueller et al.*, 1969 c; *Otten et al.*, 1973 a).

d) The increase in the activity of enzyme preparations corresponds to the increase in competition for binding sites of specific antibodies to tyrosine hydroxylase (*Joh et al.*, 1973).

The increase in neuronally mediated enzyme induction is not a reflection of a general increase in protein synthesis. Under experimental conditions which lead to a 2 to 3-fold increase in the activity of tyrosine hydroxylase in the rat superior cervical ganglia the total protein content of this organ does not change to a measurable extent (*Mueller et al.*, 1969 b; *Otten et al.*, 1973 a) although the volume of the adrenergic cell bodies contribute at least 50 % to the total volume of the superior cervical ganglion.

However, the trans-synaptic induction is not selectively confined to tyrosine hydroxylase. It also involves dopamine-β-hydroxylase (*Molinoff et al.*, 1970; *Thoenen et al.*, 1971), another enzyme selectively located in adrenergic neurons and adrenal chromaffin cells. In contrast, the activity of the third enzyme engaged in the formation of the adrenergic transmitter, dopa decarboxylase (*Black et al.*, 1971 c; *Thoenen et al.*, 1971) is not changed as is the case for all the other enzymes studied so far which are involved in the metabolic degradation of the adrenergic transmitter or in more general neuronal function (*Molinoff et al.*, 1970).

Although the trans-synaptic induction is a slow process it has been shown that an increased activity of the preganglionic cholinergic nerves of 2 to 3 hours is sufficient to initiate the events leading to a measurable increase in tyrosine hydroxylase activity after 24 hours (*Otten et al.*, 1973 a). The maximal increase is reached after 48 hours. The regulation of the enzyme synthesis takes place at the transcription level (*Mueller et al.*, 1969 c; *Otten et al.*, 1973 a) although it cannot definitely be decided whether the regulation involves the synthesis of the messenger RNA of the enzymes themselves or of a regulatory factor specifically acting at the translation level.

The regulation at the transcription level initiated by the changes in the neuronal membrane resulting from the activity of the pre-ganglionic cholinergic fibers seems to be terminated after about 18 hours, whereas the enhanced synthesis at the translation level continues up to 48 hours. These aspects of the regulation of trans-synaptic induction have been studied in detail by *Otten et al.* (1973 a) and for more complete information the reader is referred to this publication.

Possible Mediatiors of Trans-Synaptic Induction; First and Second Messenger

The trans-synaptic induction of enzymes in the rat superior cervical ganglion cannot only be blocked by transection of the pre-ganglionic cholinergic fibers but also by administration of ganglionic blocking agents (*Mueller et al.*, 1970; *Otten et al.*, 1973 a). These findings suggest that the first mediator between the preganglionic cholinergic fibers and the neuronal membrane of the terminal adrenergic neuron is acetylcholine rather than another so far unknown (trophic) substance liberated by nerve impulses from the cholinergic nerve terminals.

In view of the wide-spread function of cyclic AMP as a second messenger in many hormonal and neurohumoral systems (cf. *Robinson et al.*, 1968; *Pastan* and *Perlman*, 1971; *Rall*, 1972), this cyclic nucleotide was an obvious candidate to look at as a second messenger. Indeed, there are several observations which could be taken to indicate that cyclic AMP acts as a second messenger in trans-synaptic enzyme induction.

a) High concentrations of dibutyryl cyclic AMP produce an increase in tyrosine hydroxylase and dopamine-β-hydroxylase activity in mouse and rat sympathetic ganglia kept in organ culture (*Mac Kay* and *Iversen*, 1972; *Keen* and *McLean*, 1972).

b) Under particular experimental conditions there is a correlation between the rate of increase in cyclic AMP in the adrenal medulla and the subsequent induction of tyrosine hydroxylase (*Costa* and *Guidotti*, 1973; *Guidotti* and *Costa*, 1973).

However, there are a series of very recent observations which cast severe doubts on the correctness of the conclusions that the findings mentioned above can be taken as relevant arguments in favor of cyclic AMP acting as a second messenger.

1. *Goodman et al.* (1974) have shown that in mouse sympathetic ganglia kept in organ culture high concentrations of dibutyryl cyclic AMP produce not only a cycloheximide-sensitive increase in tyrosine hydroxylase and dopamine-β-hydroxylase but also a similar increase in dopa decarboxylase and monoamine oxidase. The two latter enzymes have been shown to remain unchanged (*Molinoff et al.*, 1970; *Black et al.*, 1971 c; *Thoenen et al.*, 1971; *Otten et al.*, 1973 a) under experimental conditions which lead to a trans-synaptic induction of tyrosine hydroxylase and dopamine-β-hydroxylase.

2. It has been shown under various experimental conditions that both in the superior cervical ganglion and the adrenal medulla it is possible to dissociate the changes in cyclic AMP from the subsequent induction of tyrosine hydroxylase *i.e.* tyrosine hydroxylase induction can occur without preceeding increase in the level of cyclic AMP and marked, rapid rises in cyclic AMP are not followed by an increase in tyrosine hydroxylase. These aspects of trans-synaptic enzyme induction have been discussed extensively in recent publications and for detailed information the reader is referred to them (*Otten et al.*, 1973 b, 1974).

As to the possible involvement of cyclic GMP in trans-synaptic induction the available information does not yet allow definite conclusions. However, in all the systems studied so far in which acetylcholine activates guanyl cyclase this activation occurs by a muscarinic mechanism (*Lee et al.*, 1972; *George et al.*, 1973) whereas for trans-synaptic induction nicotinic receptors seem to be involved (*Mueller et al.*, 1970; *Otten et al.*, 1973 a).

Concluding Remarks

The regulation of the synthesis of macromolecules by nerve impulses represents the basis for the property of integrated neuronal systems to adapt to changing functional requirements. This property represents a fundamental difference to the function of a computer. The response of a neuron to a chemically mediated message from another neuron is not confined to actual effects manifested by electrical phenomena or changes in the intermediate metabolism. The response also involves changes in the regulation of the synthesis of macromolecules which are of importance for the regulation of onto-genetic processes, adaptation to increased transmitter utilization and possibly also for the long-term storage of information in the central nervous system.

The relatively simply organized peripheral sympathetic nervous system has been shown to lend itself favorably for studying the detailed mechanism of the relationship between the changes effected in the neuronal membrane by the action of a transmitter substance liberated from the nerve terminals of another neuron and the regulation of the expression of the available genetic information.

Beyond the great importance of the regulation of the synthesis of macromolecules by nerve impulses for specific functional properties of integrated neuronal systems, the relationship between the changes taking place in the cell membrane and the regulation of the synthesis of macromolecules is of rather general biological interest. There is a remarkable similarity between trans-synaptic enzyme induction and the regulation of the synthesis of antibodies by lymphocytes in consequence of the combination of cell-specific antibodies located in the cell membrane with corresponding antigens. Furthermore, there are also similarities to the sequence of events occuring in cell culture in consequence of membrane-membrane interaction collectively referred to by the term "contact inhibition". In summary, the trans-synaptic regulation of the synthesis of neuronal macromolecules by nerve impulses represents nothing more than a well defined particular case of a general biological manifestation, *i.e.* the reaction of a cell to its environment involving the linkage between changes in the cell membrane and the regulation of the synthesis of macromolecular cell constituents.

References

Black, I. B., F. E. Bloom, I. A. Hendry, and *L. L. Iversen:* Growth and development of a sympathetic ganglion: maturation of transmitter enzymes and synapse formation in the mouse superior cervical ganglion. J. Physiol. *215,* 23p—24p (1971 a).

Black, I. B., I. A. Hendry, and *L. L. Iversen:* Transsynaptic regulation of growth and development of adrenergic neurons in a mouse sympathetic ganglion. Brain Res. *34,* 229—240 (1971 b).

Black, I. B., I. A. Hendry, and *L. L. Iversen:* Differences in the regulation of tyrosine hydroxylase and dopa decarboxylase in sympathetic ganglia and adrenals. Nature New Biol. *231,* 27—29 (1971 c).

Black, I. B., I. A. Hendry, and *L. L. Iversen:* Effects of surgical decentralization and nerve growth factor on the maturation of adrenergic neurons in a mouse sympathetic ganglion. J. Neurochem. *19,* 1367—1377 (1972).

Costa, E., and *A. Guidotti:* The role of 3′, 5′-cyclic adenosine monophosphate in the regulation of adrenal medullary function. In: New Concepts in Neurotransmitter Regulation (*Mandell, A. J.,* ed.), pp. 135 to 152. New York: Plenum Publ. Corp. 1973.

George, W. J., R. D. Wilkerson, and *Ph. J. Kadowitz:* Influence of acetylcholine on contractile force and cyclic nucleotide levels in the isolated perfused rat heart. J. Pharmacol. Exp. Ther. *184,* 228—235 (1973).

Goodman, R., F. Oesch, and *H. Thoenen:* Changes in enzyme patterns produced by potassium depolarization and dibutyryl cyclic AMP in organ culture of sympathetic ganglia. J. Neurochem. (in press, 1974).

Guidotti, A., and *E. Costa:* Involvement of adenosine 3′, 5′-monophosphate in the activation of tyrosine hydroxylase elicited by drugs. Science *179,* 902—904 (1973).

Joh, T. H., C. Gegliman, and *D. J. Reis:* Immunochemical demonstration of increased accumulation of tyrosine hydroxylase protein in sympathetic ganglia and adrenal medulla elicited by reserpine. Proc. Nat. Acad. Sci. (U.S.A.) *70,* 2767—2771 (1973).

Keen, P., and *W. G. McLean:* Effect of dibutyryl cyclic AMP on levels of dopamine β-hydroxylase in isolated superior cervical ganglia. Naunyn-Schmiedebergs Arch. Pharmacol. *275,* 465—469 (1972).

Lee, T. P., J. F. Kuo, and *P. Greengard:* Role of muscarinic cholinergic receptors in regulation of guanosine 3′, 5′-cyclic monophosphate content in mammalian brain, heart, muscle and intestinal smooth muscle. Proc. Nat. Acad. Sci. (U.S.A.) *69,* 3287—3291 (1972).

Mac Kay, A. V. P., and *L. L. Iversen:* Increased tyrosine hydroxylase activity of sympathetic ganglia cultured in the presence of dibutyryl cyclic AMP. Brain Res. *48,* 424—426 (1972).

Molinoff, P. B., S. Brimijoin, R. Weinshilboum, and *J. Axelrod:* Neurally mediated increase in dopamine β-hydroxylase activity. Proc. Nat. Acad. Sci. (U.S.A.) *66,* 453—458 (1970).

Molinoff, P. B., and *J. Axelrod:* Biochemistry of catecholamines. Annu. Rev. Biochem. *40,* 465—500 (1971).

Mueller, R. A., H. Thoenen, and *J. Axelrod:* Adrenal tyrosine hydroxylase; compensatory increase in activity after chemical sympathectomy. Science *158,* 468—469 (1969 a).

Mueller, R. A., H. Thoenen, and *J. Axelrod:* Increase in tyrosine hydroxylase activity after reserpine administration. J. Pharmacol. Exp. Ther. *169,* 74—79 (1969 b).

Mueller, R. A., H. Thoenen, and *J. Axelrod:* Inhibition of trans-synaptically increased tyrosine hydroxylase activity by cycloheximide and actinomycin D. Mol. Pharmacol. *5,* 463—469 (1969 c).

Mueller, R. A., H. Thoenen, and *J. Axelrod:* Inhibition of neuronally induced tyrosine hydroxylase by nicotinic receptor blockade. Europ. J. Pharmacol. *10,* 51—56 (1970).

Otten, U., U. Paravicini, F. Oesch, and *H. Thoenen:* Time requirement for the single steps of trans-synaptic induction of tyrosine hydroxylase in the peripheral sympathetic nervous system. Naunyn-Schmiedebergs Arch. Pharmacol. *280,* 117—127 (1973 a).

Otten, U., F. Oesch, and *H. Thoenen:* Dissociation between changes in cyclic AMP and subsequent induction of TH in the rat superior cervical ganglion and adrenal medulla. Naunyn-Schmiedeberg Arch. Pharmacol. *280,* 129—140 (1973 b).

Otten, U., R. A. Mueller, F. Oesch, and *H. Thoenen:* Location of an iso-proterenol-responsive cyclic AMP-pool in adrenergic nerve cell bodies and its relationship to tyrosine hydroxylase induction. Proc. Nat. Acad. Sci., U.S.A. (in press, 1974).

Pastan, I., and *R. C. Perlman:* Cyclic AMP in metabolism. Nature New Biol. *229,* 5—7 (1971).

Rall, Th. W.: Role of adenosine 3′, 5′-monophosphate (cyclic AMP) in action of catecholamines. Pharmacol. Rev. *24,* 399—409 (1972).

Richter, D.: Protein metabolism and functional activity. In: Protein Metabolism of the Nervous System (*Lajtha, A.,* ed.), pp. 241—258. New York: Plenum Press. 1970.

Robinson, G. A., R. W. Butcher, and *E. W. Sutherland:* Cyclic AMP. Annu. Rev. Biochem. *37,* 149—174 (1968).

Thoenen, H., R. A. Mueller, and *J. Axelrod:* Increased tyrosine hydroxylase activity after drug-induced alteration of sympathetic transmission. Nature *221,* 1264 (1969).

Thoenen, H., R. Kettler, W. Burkhard, and *A. Saner:* Neuronally mediated control of enzymes involved in the synthesis of norepinephrine; are they regulated as an operational unit? Naunyn-Schmiedebergs Arch. Pharmacol. *270,* 146—160 (1971).

Thoenen, H.: Neuronally mediated enzyme induction in adrenergic neurons and adrenal chromaffin cells. Biochem. Soc. Symp. *36,* 3—15 (1972 a).

Thoenen, H.: Comparison between the effect of neuronal activity and nerve growth factor on enzymes involved in the synthesis of norepinephrine. Pharmacol. Rev. *24,* 255—267 (1972 b).

Thoenen, H., and *F. Oesch:* New enzyme synthesis as a long-term adaptation to increased transmitter utilization. In: New Concepts in Neurotransmitter Regulation (*Mandell, A. J.,* ed.), pp. 33—51. New York: Plenum Publ. Corp. 1973.

Weiner, N., G. Cloutier, R. Bjur, and *R. I. Pfeffer:* Modification of nor-epinephrine synthesis in intact tissue by drugs and during short-term adrenergic nerve stimulation. Pharmacol. Rev. *24,* 203—222 (1972).

Author's address: Prof. Dr. *H. Thoenen,* Department of Pharmacology, Biocenter of the University, Klingelbergstraße 70, CH-4056 Basel, Switzerland.

Discussion

Wernig: Have you excluded the possibility that it is the activity of the postsynaptic cell itself rather than a trophic influence of the presynaptic cell which triggers postsynaptic enzyme induction.

Thoenen: It seems that for the trans-synaptic induction of tyrosine hydroxylase in sympathetic ganglia the nicotinic (excitatory) effect of acetylcholine is responsible. However, it cannot be concluded whether the depolarization as such (and the cascade of events evolving from it) or another nicotinic mechanism not related to depolarization is responsible. To get more information on this aspect, it would be necessary to evaluate the effect of retrogradely propagated nerve impulses resulting from electrical stimulation of the postganglionic adrenergic fibers. However, it seems to be very questionable whether meaningful results would be obtained for technical reasons. For instance, the stimulation of the postganglionic adrenergic fibers of the rat superior cervical ganglion without impairing the blood supply to the ganglionic cell bodies is technically very difficult. Furthermore, the stimulation has to be performed in an identical way in a relatively large number of animals which must then survive for at least 24 hours after termination of stimulation.

Zenker: Could the effects you showed in nervous tissue be also persumed as existent in reactive hypertrophy of muscles?

Thoenen: It seems to be reasonable that trans-synaptic regulation of the synthesis of macromolecules is not confined to the interneuronal level but occurs also at the neuron-effector cell level. I am not aware of studies designed to elucidate the detailed mechanisms linking the changes effected by acetylcholine in the muscular cell membrane and the subsequent hypertrophy. However, the relationship between the activity of the motor neuron and the extent of the sensitivity of the muscular cell membrane to acetylcholine has been studied very carefully. It seems that the expanded area occupied by reactive acetylcholine receptors after transection of the motor nerves (and consecutive degeneration of the motor nerve terminals) does not result from the absence of quantal release of acetylcholine. This can be deduced from recent experiments of *Thesleff et al.* They have shown that after blockade of the proximo-distal axonal transport in the motor neuron with colchicine there was a spreading out of the sensitivity to acetylcholine beyond the area of the motor endplate. However, the size and pattern of both miniature and evoked endplate potentials were still normal.

Pilgrim: Could it be that, besides an increase of the synthesis rate of the enzyme TH, an increase of the degradation rate also occurs and that this accounts for the long lag phase before the appearance of a measurable activity increase?

What is known about the turnover of the enzymes which you investigated?

Thoenen: As to your first question I would like to say that this possibility certainly cannot be excluded but that this question is also a rather scholastic one. I cannot think of an experimental procedure to settle this question unless you assume that tyrosine hydroxylase (TH) is separated into two different pools, *i.e.* one containing the newly synthesized TH (not available for degradation) and the other containing the "old" TH (not or very slowly mixing with the newly synthesized) and also assume that the "old" TH exclusively is available for degradation. Furthermore, one would have to assume that the induction of TH and the degrading enzyme are synchronous with respect to time and rate regardless of whether the stimulation is short or long, strong or relatively weak, since the delayed increase in TH has been observed under any experimental conditions leading to trans-synaptic induction. Moreover, it has also to be born in mind that after an increased preganglionic activity of 2—3 hours the completion of the transcription phase takes at least 18 hours. The completion of the transcription phase can be impaired between the termination of the stimulation and 18 hours and this inpairement is inversely proportional to the elapsed time prior to administration of the inhibitor. This would mean that the transcriptional and translational regulation of TH and degrading enzyme must be perfectly synchronized in the first 18 hours and then must lose this synchronization later on.

As to the turnover of tyrosine hydroxylase, dopamine β-hydroxylase and dopa decarboxylase, the data available so far were obtained by measuring the rate of decay of enzyme activity after inhibition of protein synthesis with cycloheximide at a dosage schedule providing a 95 % reduction of ^{3}H-leucine incorporation into protein. For TH the rate of decay was too small to be detectable during 12 hours of inhibition of protein synthesis (the longest time of survival of rats treated with this dosage schedule). For dopamine β-hydroxylase the t 1/2 amounted to 13 hours, for dopa decarboxylase to 12 hours. We are quite aware of the considerable drawbacks of this procedure, particularly of the fact that cycloheximide may not only impair the synthesis of the enzymes investigated but also of the proteolytic enzymes responsible for their degradation.

Seitelberger: Do you think that there are principal relations between the processes you demonstrated so impressively and the central nervous processes connected with learning and memory? What is your opinion about the so-called memory substances, *e.g.* the Scotophobin peptide found by *Ungar?*

Thoenen: It would well be that for long-term storage of information in the brain similar mechanisms are involved as in trans-synaptic enzyme induction of sympathetic ganglia. It may be that in long-term storage of information in the brain the regulation of the synthesis of synaptic structural elements are involved providing changes in their functional properties enabling the maintenance of structural changes. As to the "memory substance" scotophobin it seems that the situation is rather doubtful and I would like to refer you to the recent critical review of the scotophobin story by *Avram Goldstein* [Nature *242*, 60 (1973)].

Author Index

Subject Index